Song of the Hammer and Drill

The Colorado San Juans, 1860-1914

Song of the Hammer and Drill

The Colorado San Juans, 1860-1914

Duane A. Smith

Colorado School of Mines Press, Golden, Colorado

OTHER CSM PRESS PUBLICATIONS ON MINING HISTORY

Fading Shadows, L.W. LeRoy and J.J. Finney, 1973.
Secure the Shadow, Duane A. Smith and Hank Wieler, 1980.

©1982 Colorado School of Mines
All rights reserved
Guy T. McBride, Jr., President
Colorado School of Mines, Golden, Colorado 80401
Printed in the United States of America
Published by Colorado School of Mines Press,
Publications Center, Golden, Colorado 80401

Library of Congress Cataloging in Publication Data

Smith, Duane A.
    Song of the hammer and drill.

  Bibliography: p.
  Includes index.
  1. Mines and mineral resources—San Juan Mountains Region
(Colo. and N.M.)—History. 2. Miners—San Juan Mountains Region
(Colo. and N.M.)—History. I. Title. II. Title: Colorado San Juans,
1860-1914.

TN25.S26S64	338.2'09788'38	82-4304
ISBN 0-918062-49-7		AACR2

To Evelyn and Seth Woodruff

Contents

Illustrations

Preface

The following inscription appears on a gravestone in the Rico cemetery: "Resting amidst the hills and the people that she loved." This epitaph could well serve the San Juaners, known and unknown, who opened and developed this mountainous land from 1860 to World War I.

The San Juans were, and are, a great mining region. Thomas Rickard, well-known mining engineer and Colorado State Geologist, called them in 1896 one of Colorado's four greatest mining regions. It took much perseverance, toil, money, and heartbreak to attain this distinction. Regarding early La Plata Canyon development, Rickard cagily observed, "great expectations had a sequel of small accomplishment." Forty years later, a distressed investor lamented: "Every now and then I get heartily sick of the mining business. . . . It is a hell of a business and I get sick of sitting here listening to stories, ninety-nine percent of which I am convinced are the bunk." It was all this—but also fascinating, exciting, and rewarding for the fortunate few.

The purpose of this volume is to study the "hills" (the mines) and the "people" who worked them and lived in the camps. It is the story of ordinary people doing ordinary things to settle southwestern Colorado. For them it was a work-a-day, live-a-day world; they would be amazed, perhaps amused and pleased, that their efforts have acquired legendary status.

The references appearing at the end of each chapter serve as a bibliography and source for further research, for those so inclined.

I owe an unpayable debt of thanks to many individuals for their help in the search for the San Juaners and their history. Allan Bird and Orval Jahnke, twentieth-century San Juan mining engineers, assisted in many ways. Mrs. Marvin Gregory, Mrs. Homer Reid, and Charles Engel provided photographs. The staffs of the Colorado Historical Society, Henry E. Huntington Library, Durango Public Library, San Juan County Historical Society, Western History Department of the Denver Public Library, Western Historical Collections of the University of Colorado, American Heritage Center—University of Wyoming, and the Fort Lewis College Library and Center of Southwest Studies provided their usual courteous and professional assistance. The city clerks of Rico, Telluride, Ouray, Silverton, Durango, Creede, and Lake City have kindly allowed me to research the city records over the past decade. Many oldtimers and others shared reminiscences and information; to all of them my sincere gratitude.

I also wish to acknowledge the generous financial support of the American Philosophical Society, Huntington Library, and University of Colorado Centennial Commission. The help they provided allowed me to research for two summers and purchase photographs. My sincere appreciation to Jon Raese, who shepherded the manuscript into the printed page. Thanks alone cannot express my gratitude to my wife, who labored long over this volume; without her support and help, it would never have become reality.

Finally, no words can adequately convey my respect for the San Juaners themselves, whose adventures made this volume possible. I, too, love the "hills and the people" and hope this will in a small way pay tribute to them, their era, and their significance.

This book is dedicated to Evelyn and the late Seth Woodruff, always loyal in their support and everything a son-in-law could wish.

Prologue

The Magnet of Mining

The San Juans—the majestic San Juans—words do them no justice. To comprehend fully their beauty one needs to hike and climb, to stretch out among the peaks; then and only then does one begin to savor the towering mountains and the alpine meadows. A land of contrasts, this, of wildness and sublime beauty, of man and of nature. Thunder rolling over the mountains and cascading down the deep canyons to ebb away in distant valleys, the still silence of a snow-crested meadow, the awesome grandeur of rugged 14,000-foot peaks—all are there to be seen and felt, to wonder about and respect.

In the past centuries, these mountains have attracted the gold and silver seekers. Man came, leaving behind scars, legends, and frustrated dreams; the mountains stayed, changing little. They still stand nearly as the 1876er saw them; he who came to match the San Juans could not know what they held in store.

Over the years there has been some controversy over the definition of the territory that constitutes the San Juan country, a controversy generally resolved in the 1880s. One of the original San Juan boosters, William Weston, writing in the *Engineering and Mining Journal* in April 1882, outlined the region as Ouray, Dolores, Hinsdale, La Plata, San Juan, and Rio Grande counties. Add Mineral and San Miguel counties, both carved out of older counties after 1882, and the San Juans as defined in this volume are introduced.

Tucked away within these man-made geographical boundaries are the San Juan Mountains, named long ago by the Spanish, and the headwaters of numerous rivers, including the Rio Grande. It is a land of high valleys and rocky canyons, topped by nearly a score of 13,000 to 14,000 foot peaks, with probably the highest average elevation in the United States, over 10,400 feet. The mining districts of these eight counties cover roughly 4,891 square miles. Because of their isolation and mountainous terrain, they remained one of the last sections of Colorado to be opened to mineral exploitation and permanent settlement. Man challenged this region for better than a century before a permanent toehold was established in the 1870s, and then nearly two decades passed before the mines came into their own on the Colorado and national mining scene.

This is the story of the San Juans from the first American mining penetration in 1860 until the end of 1914, when World War I brought so many changes to this region and the whole country. It is the story of the growth of a mining economy that was forced to turn to eastern and European capital and technology when western expectations exceeded resources. The San Juaners found that such an expedient placed them in something approximating colonial status, with mixed blessings for all concerned. They should not have been confounded at this result—individualism invariably gave way to corporation dominance on the mining frontier, as did economic independence to dependence.

During the fifty-four years under discussion, a regional economy emerged, bulwarked by mining, shored by outside finances. The prospectors and miners first sought gold and it was there, an estimated $108,066,000 by the close of 1914. For a while these were also called the silvery San Juans, with good reason; $106,136,000 in silver was mined, achieving very nearly a production balance between these two precious metals. Add to these totals those for the other minerals and coal, and the significance of this region emerges. Only the far better known Leadville and Cripple Creek districts in Colorado and a few outside the state produced more during these years.

Mining never stood alone. Farming and ranching soon appeared, along with lumbering, transportation, and a general business community, all attracted by the lucrative market. Early attempts were made to capitalize on the natural beauty and attract the tourist, who foreshadowed later developments and a different regional economic base. The San Juan economy progressed sputteringly with fits and starts, suffering from national dislocations, its own youth, and specialization. Eventually, all meshed together and the dreams of the pioneers were realized, far beyond what they had imagined. For the San Juaners during most of these years, the guiding stars of their destiny were optimism, faith, and determination. Not until near the end of these decades did these begin to fade and be replaced by an underlying insecurity.

A mining economy is inherently urban because of the specialization involved and the availability of money to buy services. As a result, the San Juans urbanized to an extent that has not been matched since. The little camps, some of which grew into towns, were the epitome of optimism and as such strove to imitate midwestern and eastern contemporaries. They did so with a speed seldom matched elsewhere in America; visitors were amazed and townspeople proud of the growth and development which, within a season, transformed the wilderness to a roughhewn settlement, then a thriving mining community. Anxiety lurked beneath a calm exterior; community leaders desperately tried to build their economic base and outrace competitors, even to the extent of siphoning off rival towns' residents to build up their own. They attempted to widen their business, social, and political clientele and draw smaller camps into their economic orbit. A few succeeded; the rest died or declined. Before these fifty-four years passed, nearly all the camps had peaked in population, wealth, and power, and many had become only a memory of a mining community, whose season of life had withered. Still the process continues; Ironton, according to the Federal Census Bureau, reached ghost-town status in 1970.

During the noonday of their existence, these communities were home for a generation of San Juaners who rode the crest of exploitive mining, if not to fortune, at least to income sufficient to meet their needs, with enough left over for some good times. Towns stood ready to provide a variety of services, centering upon what was understood to be culture. Refined visitors might hesitate to call it such, but to the San Juaners it was all the embellishments that money could buy.

They came by the hundreds, then by the thousands, always searching, forever seeking a dream. Unsung and unknown today, they labored, played, succeeded, and failed. Almost all spent their lives in obscurity; indeed, the San Juans were hardly known outside Colorado until the late 1890s. The Rev. James Gibbons, who ministered to many of them, concluded in his book, *In the San Juan* (published in 1898): "Life in the mining regions, especially the wilds of the San Juan, is little known to eastern people." To provide recognition for these people, to shed some light on their lives, is the purpose of this study. Much of what they did individually has been irretrievably lost, yet enough is left to reconstruct and breathe life into the dry facts of history.

David Lavender, in his novel *Red Mountain*, has one of his old-time San Juaners answer an easterner's question as to why the people settled and built Red Mountain (the fictional name for Ironton). Musing for a moment, the man reviewed such possibilities as hard work, perseverance, courage, self-reliance, ingenuity, and greed. "What made them go was a sort of urge, a frame of mind. . . . That's what built Red Mountain—the frame of mind of the people. . . . What frame? They believed." Believe they did—that the San Juans would be their salvation. They had a faith that allowed them to stand against an uncompassionate envir-

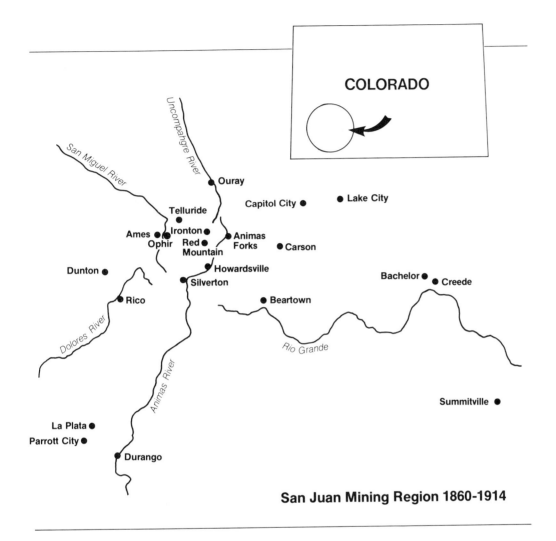

COLORADO

San Miguel River
Uncompahgre River

● Ouray
Capitol City ●   ● Lake City
Telluride
●
Ames ● ● Ironton ●   ● Animas
Ophir   Red ●   Forks
Mountain   ● Carson
Dunton ●
● Howardsville
● Silverton
● Rico   ● Beartown   Bachelor ●
Dolores River   ● Creede
Animas River
Rio Grande

● Summitville

La Plata ●
Parrott City ●
● Durango

San Juan Mining Region 1860-1914

onment and compete with their fellow man, to persevere against long odds, to build a life where everything conspired against them. As such they were heirs to an American frontier tradition. John Winthrop and his Puritans would have understood; so would Daniel Boone and the 1840s Oregon pioneers.

Insofar as possible, they will herein tell their own story. At this late date, no one can fully place himself into their lives. The times are gone; the sights, the sounds, the smells have vanished. Some of their communities stand yet; the big six (Telluride, Silverton, Ouray, Creede, Lake City, and Rico) exist only as pale shadows of yesterday's glories. The mines they worked continue to produce; 100 years is an incredibly long time to mine in one district. In the West only a handful of their contemporaries can claim the same record. Two counties, Mineral and San Juan, were the number one gold and silver producers, respectively, in Colorado in 1980. Between them they produced over $40 million, having surpassed their long-ago rivals of Gilpin, Lake, and Teller counties. This treasure chest produced an abundance of other minerals as well, including copper, vanadium, lead, zinc, uranium, and coal.

Fate was not always kind to those who risked their lives and fortunes in the dark depths of the San Juans. Blinded by a premature explosion in March 1900, miner Alfred King knew how suddenly fortunes could change. As tragic as this accident was personally, it turned his attention to writing poetry "from the recesses of memory," recalling his occupation and era. From King's poem "The Miner," a tribute to the San Juaners, comes the title of this book:

Clink! Clink! Clink!
The song of the hammer and drill:
At the sound of the whistle so shrill and clear,
He must leave the wife and the children dear,
In his cabin upon the hill.

Clink! Clink! Clink!
But the arms that deliver the sturdy stroke,
Ere the shift is done, may be crushed or broke,
Or the life may succumb to the gas and smoke,
Which underground caverns fill.

Clink! Clink! Clink!
The song of the hammer and drill:
As he toils in the shaft, in the stope or raise,
'Mid dangers which lurk, but elude the gaze,
His nerves with no terrors thrill.

Clink! Clink! Clink!
For the heart of the miner is strong and brave;
Though the rocks may fall, and the shaft may cave
And become his dungeon, if not his grave,
He braves every thought of ill.

G o slow," warned William Byers in the  *Rocky Mountain News*, November 9, 1860; "very many of our people are becoming excited on the reported richness and promise of the San Juan mines." The Pike's Peak country's foremost newspaperman went on, "It is evident that there is 'speculation and profit' in the enterprise for a few who are already interested in these new discoveries; but nothing tangible and authentic has reached us of sufficient weight to warrant the rush to that mythical *dorado*." Thus the bewitching San Juans made their debut on the public scene amid controversy and doubt. Mythical? Only in the sense that men had long dreamed of mineral wealth there, and a very few had already measured that dream against those jagged mountains.

"The Mythical Dorado"

Byers was honestly concerned about the rumors of "great mineral wealth," as a letter the previous month had proclaimed. In 1860, when Colorado was young and mining new, anything seemed possible; that the "gold belt" passing through Boulder, Central City, and California Gulch could extend southwesterly was not inconceivable. Who would argue with the writer who claimed that "in all probability [it] grows richer as one goes in that direction?" Byers would and did. He was concerned about the attempt to enter the "great snowy range" so late in the season; even then the winters in this basically unknown area evoked awe. Byers fretted even more when Santa Fe, New Mexico, and Canon City each claimed to be the gateway, stealing business and headlines from his Denver. Finally, there remained the possible aftermath, if this turned out to be a hoax, "a hasty, rash and ill-advised venture." Byers knew that the 1859 rush had almost turned out to be just that, and he did not want a repeat performance to blemish Colorado's name (which name the territory did not officially have as yet).

This furor could be traced to Charles Baker, who in August 1860 led a small party into what became known as Baker's Park. By that mysterious, always buzzing mining camp grapevine, hints of a bonanza spread. Increasingly favorable reports gave only vague notions of location, however. What specifically lured the "restless, adventurous, impecunious" Baker into the heart of the San Juans remains unknown.

Without adequate information, moving toward an unknown destination, the vanguard of miners set out that fall. Newspaper reports from Santa Fe fanned the flames with some specifics. The old village of Abiquiu on the Chama River (a "miserable village" in a "God forsaken land" moaned a Denverite) was the nearest settlement to the San Juans. From it stories originated regarding Indian agent Albert Pfeiffer's tour, during which gold had been found. The news created interest wherever it went. Canon City, erstwhile gateway, was nearly depopulated, as men left for the "San Wan" mountains. Toward Abiquiu the hopeful journeyed; many wintered there, organizing themselves into companies and making preparations to "pan clean" the snow-locked San Juans.

One Denver party of this period, heading for the "gold land," crossed the San Luis Valley from Fort Garland aiming for the Uncompahgre River. They made it, despite the snow, and wintered near the future site of Ouray. This group of men nearly starved, being reduced to "very slight" rations before they could work their way out in the spring, *sans* gold. For them the San Juans held only cold and snow; the gold was still locked in their dreams.

Ore samples and more letters arrived in Denver, forcing Byers to hedge a little and confess frankly that he did not know what to believe. Others were not so hesitant. By mid-January 1861, a weekly express messenger left Denver for Taos, and merchants were moving to supply the expected spring rush.[1]

Finally, Byers found what he was seeking in his attempts to bring order out of the chaos of contradictory rumors: interviews with men who had been there, in a letter from a trusted correspondent. The picture they presented proved discouraging. Byers' sources accused New Mexico traders of fabricating the excitement in order to sell a surplus of merchandise and reverse the dull times settling on that territory. The men who had been there told this tale: after a "desperate effort," they reached deserted Baker's Park, finding there a "large number of prospect holes and several gulches staked off." Their panning proved unrewarding. The best pan produced $2.50 in gold, most only a few flakes—scant reward for the hard labor, in a country an "old experienced miner" said did not "look right for gold." Enough was enough; they raced to escape before winter, "the worse used up and most ragged men that I ever saw." These reports should have stopped the rush, but logic has little appeal in a brewing mining excitement.

Almost providentially, a letter from Baker arrived which described his activities in some detail. According to the discoverer, the gulches and bar diggings were rich, "richer than any mines hitherto discovered to the North-East of them," and of sufficient quantity to give "profitable employment" to all who came. Men had already gone to search the "San Meguil [sic] river and to the Dolores and the Rio de los Mancus [sic]." In early October 1860, Baker and his party reached Abiquiu and immediately started constructing a road over the 175 miles from there to the mines. Before leaving the San Juans, they had organized a mining district and claimed a town site. All these activities were familiar ones on the mining frontier, and Baker's group obviously planned to profit from everyone who came after them. Animas City, over the mountains and to the south of the Park, had already been organized in the river valley, and 300 to 500 men had settled down to wait for spring. Baker predicted that no fewer than 25,000 Americans would be in the region the next year to mine and farm.[2]

Despite his earlier reluctance, even Byers started to capitulate, especially when Denver's rivals got the jump on his beloved city. Was not Denver nearer the Missouri River ports and would not a good road soon be built to the mines? He thought so. Byers refused to succumb completely, though, reminding his readers of other mining stampedes in California, and of the one to Tarryall in 1859, when the stampeders did not stop in Denver on their way back, but "rushed on to the States." God forbid such a humbug should happen again. God, too, was on his way there; the Methodists were even then considering sending a circuit rider to the San Juan mines.

As spring approached, the first stampeders did return with disheartening experiences. One returnee, thinking Baker still might find good diggings, nevertheless bluntly emphasized that the previous reports had been grossly exaggerated. Another claimed Baker said, "I alone know where the best dirt is." While Byers fretted over the continued dribble of reliable news, the men at Abiquiu fared no better. The 300 to 500 at Animas City (if that were not a figment of Baker's imagination) shrank to seventy-five beleaguered residents, who found the cost of living high and the snow deep.[3] Still, the optimistic felt that the coming months would demonstrate the richness of the region.

The change of season brought more exciting news from another quarter: Civil War had broken out back in the states, and it proceeded to push the San Juans out of the headlines. Down in those mountains, which isolation made one of the last places in the continental United States to hear the news of the war, miners moved into the mountains to prove or disprove the golden stories.

Frank Hall, quoting Byers in his *History of Colorado*, gives one of the best firsthand accounts of what happened; Byers originally acquired the story from Samuel Kellogg. Kellogg, who had helped grubstake Baker, left Denver with his party in December 1860. After enduring inclement weather and arduous travel they reached Abiquiu, where they turned up the Chama River, passed what became Pagosa Springs (a planned townsite was staked there), and moved into the Animas Valley. In the party was future Colorado governor Ben Eaton, who, after his experiences, gave up mining and turned to farming, where he made a fortune. He was one of the shrewder ones. The group camped in the Cascade Creek area, while Kellogg and several others went over the mountains to Baker's Park in search of the man who gave the park its name. That much-maligned individual had been at Abiquiu earlier, where he still talked confidently but denied authorship of the stories then circulating. Baker's reputation was not enhanced by his offhand comment that he himself knew little of prospecting and had left it to others in his party. This created bitter feelings toward him and even provoked threats of violence.

Kellogg finally found the men he sought and was taken to the site of the "profitable" diggings, where, somewhat more than a decade hence, Eureka would be established. After several weeks, the best he could do—and he was an experienced prospector—was fifty cents a pan. What a pittance as reward for coming hundreds of miles! Kellogg admitted later that he and the others searched only for gulch or placer gold, knowing nothing about lodes or quartz veins and thereby missing the silver deposits. Another miner there about the same time gave a similar report; he prospected nearby creeks and the discovery site with no success. His estimate of 800 to 1,000 men in the area may be high but illustrates the attraction. He even mentioned that eight or ten families had made it into the park, showing the pioneering spirit of the women. After a few days, he had had enough and departed, but not before hearing rumors that the miners were "talking of bringing Mr. Baker to justice." Baker was not brought to justice, although the first San Juan rush proved to be the humbug Byers, among others, feared.

Upon returning to Denver, George Gregory emphatically declared that he would rather travel to the Missouri River and back six times than go again to "Animos" City and back once. Baker, he concluded, "was near a maniac as anything I can compare him to . . ." He added this interesting bit of information: Baker insisted his friends go "into the country where there was nothing, and as I believe to lead people over the Toll Road which he is interested in, and build up the towns they have located."[4]

Some diehard miners lingered in Baker's Park all summer; eventually they left, too, and everything was abandoned—the Park, Animas City, and the dreams of those who had staked so much on the San Juans. The San Juans suffered a cruel blow from which they did not recover for almost a decade. The rush left scars on the land, broken sluice boxes, deserted cabins, and a host of other odds and ends to mark the presence of men. Nothing had been proven except that Baker's stories were exaggerated; the dream of mineral riches did not die.

The gold fever had failed to pan out for a variety of reasons, the overriding one being a lack of rich diggings. Even if the placers had proven lucrative, the very isolation of the spot and the Indian threat, compounded by the withdrawal of troops because of the Civil War, weighed heavily against immediate development. Shortages of supplies and actual starvation faced those who braved the inhospitable San Juans and their changeable weather. The story would have been the same had the miners recognized the district's silver veins; Colorado was not ready to handle silver smelting.

The key to opening the San Juans, then, as it would be for decades, was transportation; without it little could be accomplished and other districts would overshadow this one. Who can say what motivated individuals to stampede to the San Juans. Public attitude was con-

"Ho, for the San Juans" was the cry. Alfred P. Camp, at right, and two friends leave Del Norte, the early gateway. All supplies had to be taken along.

The rugged San Juans as they appeared in this 1875 William Henry Jackson photograph. Jackson was standing on Solomon Mountain, viewing the future Silverton area. Prospecting and mining in this high terrain was literally breathtaking.

ducive to this type of hysteria, because all rumors seemed plausible in those intoxicating days. Perhaps the New Mexico merchants had planted stories, based at least partly on first-hand information. Baker's activities enticed others to examine the region, and the hearsay of rich discoveries whetted interest. Always in the background was the speculation, rampant since the Spanish period, that rich mineral deposits existed. The participants paid the price when this classic mining stampede busted, but what a fortune might have been theirs had it been true!

They were not alone in their excitement; similar rumors had motivated the Spanish for more than a century. Juan de Rivera led a party into the La Plata Mountains in 1765 looking for silver, supposedly guided by a Ute who knew where it was to be found. In 1776, while the thirteen colonies struggled for their independence, the Escalante expedition crossed the Animas River south of present Durango and moved west to the La Plata River. Here, on August 9, Escalante recorded in his journal, "They say there are veins and outcroppings of metal. . . . The opinion formed previously by some persons from the accounts of various Indians and of some citizens of this kingdom that they were silver mines caused the mountain to be called Sierra de la Plata."

There is no doubt that the Spanish mined in the San Juans; early Anglo miners found evidence of ancient workings with tools and equipment. The Spanish left no records, probably to avoid paying the royal fifth to the King's treasury. Undoubtedly they mined some placer pockets, perhaps a little silver ore on a seasonal basis, but the climate, isolation, and Indian pressure precluded any permanence. In the end they left behind Spanish names and legends of lost mines which still entrap the unwary into futile searches.

By the time of the American conquest of New Mexico, trappers had added a few legends of their own to the growing San Juan mystique. In 1859 Lt. John Macomb, whose exploring party skirted the southern San Juans, confessed that the name La Plata seemed to indicate silver had been found, but he could not find any definite knowledge that would justify the title. The editor of the *Santa Fe Weekly Gazette* of February 5, 1859 was more optimistic; basing his statement on unidentified "considerable observation," he said that New Mexico was full of gold. Rumors, always rumors—the San Juans were being rumored to death.

Following the collapse of the 1860-61 rush, only a few small parties ventured into the San Juans. Scant success rewarded their efforts. The determined Baker returned in 1867, after serving in the Confederate Army, only to be either killed by Indians or his companions. The Utes were granted the whole area in 1868 as part of their reservation, further compounding the difficulties. In the words of Colorado historian Frank Fossett, the San Juans remained "terra incognita," as did almost all of the Western Slope. Others besides Baker kept faith. Joel Whitney, a tireless mining booster, predicted in mid-decade that silver and gold would be found en masse in the La Platas and San Juans, echoing what the first territorial governor, William Gilpin, had said earlier. More significantly, Randolph Marcy, Inspector General of the Army, on a tour of New Mexico and the Ute country, wrote in August 1867, "It is said that there are rich gold mines in the La Plata Mountains near the Animas River, and that they cannot be worked without protection of troops as the Utes who range west of the locality are hostile."[5]

As the decade closed, the San Juans generated renewed interest. In 1869 prospectors reached the Mancos and Dolores river valleys , several working at the site of future Rico. The next year the Dolores river valley and Baker's Park were prospected. Special Agent William Arny, while visiting Colorado Indians in 1870, passed through the Animas Valley and on May 25 jotted in his journal, "On this stream are good placer gold diggings and in the mountains above are rich gold and silver quartz deposits." Even before leaving Abiquiu, he knew of three parties that had been turned back, and he talked to others who had been repulsed by the Utes. Arny sent a letter to the miners on the Dolores warning them that the Utes objected to settlements, although they were agreeable to allowing prospecting. That concession was a fatal mistake, if they wished to retain their land. Arny wisely recommended cultivating "peaceable relations" and noted in his journal that some arrangements should be made with the Utes about selling the mineral land. It was either that or remove the miners; conflict would inevitably occur if both groups remained.

In his report Arny reported that about 200 miners had been on the Ute reservation, and 274 claims had been filed on gold and silver lodes. To support his statement he included a map of the San Miguel District on the Dolores River, plus the names of claims and owners.[6] Regrettably, the Indian Bureau ignored the situation, which in turn allowed tension to heighten.

For the prospector and miner of the 1870s circumstances had changed from those of 1860-61. Colorado's transportation had improved; the railroad reached Denver in 1870 and better roads stretched into the mountains. After ten years of mining, experience and maturity replaced enthusiasm, and smelting had attained a much more scientific basis. The territorial economy had matured to the point that it could more easily support isolated communities, and Anglo settlement had reached the San Luis Valley just to the east of the San Juans.

These pioneer San Juaners did not stop to consider such improvements—they were trying to stake a claim on the "mother lode." Glittering profits tempted them to overlook the Ute threat. Gold lured them in, but silver soon caught their fancy. Colorado was entering its silver decade, a decade which would see it become the primary silver producer in the United States. The San Juans helped make this possible. Initially, however, gold stirred expectations—gold found in the streams and in a few mines. The pioneers came, staked claims, and agreed to district boundaries and enough mining laws to create some extralegal order out of their trespassing on Ute land.

Again, as in 1860, rumors spread, this time with less rapidity but more truth. A helter-skelter stampede did not develop, mostly because of the Ute barrier. The Indian Bureau tried to stem the tide, denying permission to mine on the reservation and dampening enthusiasm where it could; it might as well have shouted into the wind. By 1872 no bureaucracy fiat had proved effective. The territorial press was appalled at such efforts, dismissing completely any justification there might be. The editors of the *Daily Central City Register*, April 19, 1870, deplored the concept: "So desirable a region . . . the occupation of the mines would drive out and destroy the game, and we suppose this precarious means of support must be preserved for a few savages, to the prevention of their own civilization and the impeding of the development and progress of the American people." When this type of argument gathered momentum, the Utes were doomed. More encouraging from the miners' point of view was the fact that the Utes did not restrict their movements to any serious degree. Of course, there were not many whites yet.

The trespassers prospected and mined only for a season, returning to civilization to avoid the winter months. One of them, Donald Brown, who discovered what he called the St. Lawrence mine, went to Pueblo with high expectations: " . . . being rich in silver mines, but with only a dollar or two in my pocket. Being desirous of remaining in town for the winter, yet not being able to afford a hotel, I hunted around to see what I could find." Eventually he found a house rent free and next spring returned eagerly to his claim. Rich in dreams, poor in money—the story of the early San Juaners.

Unlike the sixties, this rush was based upon lode mining, a more stable proposition than placer operations. The Little Giant, located at the head of Arrastra Gulch and the first lode mine developed in 1870, was the favorite. Reportedly very rich —its assays ran from $1,000 to $4,000 a ton in gold—it was the first mine worked to any depth and the only one that was more than simply a name for several years. For the next two years it grabbed headlines and advertised the area. When it was sold in 1873 to Chicago investors for a reported half million dollars, the sale was heralded as the coming of age of the region. The celebrated Little Giant

The early San Juaners were prospecting on Ute land, an explosive situation. Tension stayed high, even after the mining region was ceded, as an artist for the Harpers' Weekly *displays.*

captured one more first: the first great failure. A "million dollar" mine on paper, it proved far from that in reality.

By April 1872, the rush was in full swing, with reports of three- to six-foot veins of silver, not gold, even with the Little Giant. Over rough trails they came, barely passable for man and burro. The company working the Little Giant managed to ship in a fifteen-stamp mill. Prior to this only a primitive arrastra crushed ore. Byers, whose memory remained undimmed, reminded his readers of the 1860-61 fiasco, while feeling that this time a more substantial discovery had been made. He predicted an exciting and busy season; nevertheless, having been around a long time, he correctly foresaw other aspects of the mining frontier:

> All will be bustle, hurry, noise, excitement and confusion. Doubtless there will be stores and saloons—the latter abounding—and these will be crowded with men whose pockets will be filled with big specimens, small silver bars, and rolls of location notices and assay certificates, buying, selling and talking mines; the bummers of the seediest class, who will drink—as they have for years in other camps—at the expense of every stranger who ventures beyond Sangre de Christo pass; and who can question the certainty that there will be enacted in that locality the same scenes of lawlessness which have traditionally attended such excitements, and that dance houses will be filled with half or wholly tipsy miners with, perhaps, a sprinkling of abandoned women, whose smiles will be as eagerly sought for and as jealously observed, as were ever those of that most gifted and virtuous of their sex in the abode of refinement, at Denver, St. Louis, Chicago or New York?[7]

After his early enthusiasm, Byers became more reticent and feuded with the *Tribune* over the veracity of the San Juan stories. He finally advised on August 4, 1872 that nothing warranted a stampede or even a moderate rush. This time he proved to be too cautious.

The degree of mineral richness, though fascinating, was not the pressing issue. Like a lighted powder fuse, the Indian question edged toward an explosion. The miners' encroachment could lead to a fight (the press would call it a massacre) and eventually war, which might settle the issue only at the cost of lives, money, and stability. Far better to deal with the Utes first and then have them cede the land. If there was peace, Colorado would project a sounder image in the investing East. Toward this goal Colorado Congressional delegate, urbane Jerome Chaffee, helped guide a bill through Congress to create a commission.

The commissioners arrived at the Los Pinos Agency in August, met an estimated 1,500 Utes, and got nowhere, concluding "we became satisfied that at least for the present it would be impossible to conclude any satisfactory negotiations with them." Their able agent, Charles Adams, explained why the Utes refused to modify the 1868 treaty. They "fear that should they dispose of one part, prospectors and adventurers would immediately push farther on . . . and argue that every man has the right to go where he pleases . . . ." Considering the history of white-Indian relations, this was perhaps not an unjust assumption. Ouray, the principal Ute spokesman, found time forcing his hand. The commissioners warned Washington that, although Ouray assured them trespassers would not be molested, "their continued presence on Indian territory may at any time bring about a collision."[8] The tension mounted as the issue became entangled with Colorado politics, and newspaper articles continued to boom the mines. The Utes exercised surprising restraint, in contrast to the Sioux reaction to the Black Hills invasion.

Until the question was resolved, the fate of the San Juans remained uncertain; no legal title could be given and the ever-present danger of an Indian war lurked. The government's next move threw fear into the San Juan settlers and Coloradans generally; it ordered troops to remove all intruders who had not left by June 1, 1873. Predictably, howls of protest reverberated throughout the territory and pushed Chaffee and Governor Samuel Elbert to action, who finally pressured President Ulysses Grant into suspending the obnoxious order indefinitely. Having failed to accomplish the treaty guarantees, the government had no choice but to send a second commission. Byers voiced Colorado's prevailing sentiment when he editorialized that the reservation was too large, and it was time to settle the Ute question in a manner mutually beneficial and satisfactory to both races.

This time Felix Brunot, president of the Board of Indian Commissioners and a skilled diplomat, headed the commission. The meeting started off badly when the commissioners were over two weeks late arriving, but on Saturday, September 6, 1873, the sessions opened. The Indians, having assumed from hints dropped at the previous council that all the whites wanted was the actual mining land, agreed to sell under provisions that startled Brunot. They would sell only mountain tops, none of the valleys; the miners could build no houses and would have to leave each fall. As the Utes told the commissioners, " . . . some of the miners with whom [they] came into contact, said 'the government was away in the east in the States, and had no power in the mines; it could not protect the Indians; and that they did not care whether they sold the mines or not, they were going to stay.' "[9] The canceling of the removal order, which the Utes knew about, gave support to this contention, as did the disregarding of the frequent Indian complaints about trespassing.

The council lasted for a week, with several day-long interruptions. Brunot tried to impress upon the Ute leaders that their solution could only postpone, not resolve, the impasses. He realized that the miners would not concur with any agreement he might make allowing only seasonal operation. Many related issues were discussed, including boundaries and agricultural land, while Brunot held steadily to the course of explaining every article of the agreement he brought with him and making no promises. The negotiations lasted until September 13, when the Utes finally conceded and signed what became known as the Brunot Treaty, giving up for $25,000 annually about 4,000,000 acres of mining land. One clause of the agreement gave rise to a great deal of local dissatisfaction: permission for Indians to hunt upon ceded land so long as the game lasted and peace was maintained. To satisfy some Ute leaders a tour was arranged to outline the boundaries. They also visited the mines, showing

(above) Settlement came in the river and mountain valleys. This Jackson photograph shows Baker's Park, the site of the 1860-61 excitement, and Howardsville, the first 1870s camp.

(left) Howardsville was not a year old when Jackson visited it in 1875. Despite an early start, it never grew beyond the camp stage and was soon surpassed by neighboring Silverton.

interest in them; one Indian even broke off a piece of Little Giant ore and then, with help, panned it.

Ouray, whose patience and logic endeared him to the commission, wrote Brunot what he hoped would come from the treaty:

> We are perfectly willing to sell our mountain land, and hope the miners will find heaps of gold and silver; and we have no wish to molest them or make them any trouble. We do not want they should go down into our valleys however, and kill or scare away our game.[10]

Even though the miners received virtually all they asked, the next spring Adams had to warn them about trespassing on the reservation, particularly the southern portion which closed the older New Mexico route to the mines. The Utes took the matter into their own hands, peacefully stopping all travelers. The boundary question remained a vexing one and, just as the Utes had foreseen, the whites encroached continually. Adams's successor warned in 1875 that the boundary problem still agitated "a good deal" of feeling among the Indians. The time had come, he stressed, when "there can no longer be a question as to what is whose."

For the satisfied miners all this was academic; Ute ownership was resolved and they could go on about their business. And they did, with a fervor unmatched since 1860. So aroused were they that Thomas Cree, commission secretary who accompanied the Utes on their tour, reported that just on the rumor of an agreement fifty people started to locate town lots and, following the purchase, mine owners claimed the value of their property had doubled. Since mid-1872 the population had grown slowly. Ex-Governor Gilpin generated new interest by predicting diamonds would be found, but the 300 or so men there in the fall of 1872 failed to discover any gems.

The year 1873 was the pivotal one for the San Juans; not only was the Indian question put to rest, but immigration measurably improved, and the mines became known abroad. Outside advertisement and promotion proved easier to come by than investment. Chicago papers carried San Juan stories, one calling it the "wealthiest mining district in the wide west." New York's *Engineering and Mining Journal* and San Francisco's *Mining and Scientific Press* discussed the area and its mines. The best coverage was found in the latter, which tried seriously to separate fact from fiction. The *Press* warned its readers about exaggerated assays, parochially pointing out that very few Californians mined there and the majority of the San Juaners "knew nothing at all about mining."[11] The editors did see one benefit arising from the rush: they wrote on December 13, "It will have one good effect at least, as it will cause easier times in the old camps with high wages and plenty to do." And the San Juans in this sense were a safety valve for other Colorado mining communities.

This watershed year of 1873 did not generate only blessings; problems continued to plague the miners. No smelter was available to work silver ore, nor had any method actually been tested to see what might or might not work. The trails, though showing some slight improvement, remained primitive, and transportation problems handicapped development in all respects. Nor was this what was termed a "poor man's" digging, where placer gold predominated. Paltry results hardly justified the efforts for those persisting in pan and shovel or sluice box methods. The future of the San Juans rested on quartz mining. Many claims had been staked, though few had been developed beyond scratching the surface. Quartz mining required money, skill, and time; only the last was plentiful in the San Juans. The *News* repeatedly warned those short of cash not to come. The cost of living soared nearly as high as the peaks and the demand for laborers never exceeded the supply.

Behind these problems lurked a shortage of capital. The San Juaners agreed wholeheartedly with the writer in Ecclesiastes, "Wine maketh merry; but money answereth all things." Unfortunately, despite determined efforts that included exhibiting ore at the 1873 Territorial Fair, investors were not enticed. A national economic panic, followed by a steadily worsening depression, left few people willing to gamble on an untested mining region. Unlike more established neighbors, whose mines poured forth wealth that enabled them to weather this economic storm, the young San Juans slipped into the maelstrom.

It might appear that the future looked bleak, but such was not the case. By the end of 1873 permanent settlement was assured. Already little hamlets were springing up, as the urban nature of the mining frontier exerted itself. Around the rim, La Loma and Del Norte jealously eyed each other from opposite sides of the Rio Grande, loudly proclaiming that each offered the best route, the gateway to the bonanza via unforgettable Stoney Pass. A Del Norte visitor wrote that windows were filled with gold and silver quartz and the streets crowded with enthusiastic, soon to be rich, prospectors. "We heard little but conversation on 'leads,' 'silver,' 'Baker's Park,' and 'the Animas,' and other topics all interlarded with the, to us, most incomprehensible of miners' phrases." The fever touched all; even a barkeeper mixed Little Giant sours.[12]

In the Animas Valley, where a number of miners wintered, a group founded Elbert up the valley from the site of old Animas City, and Hermosa City came into existence by the creek of the same name. The agricultural potential of the valley was also being weighed by a few unenamored with mining in the mountains. In Baker's Park, a 320-acre town site had been laid out, as yet unnamed. Some preferred "Silvertown," others "Animas City"; it finally came to be known as Howardsville. To encourage settlers, free lots were given to all who would put up "respectable" buildings. It would be, boosters claimed, the "great mining and smelting center."

Once these and other settlements evolved, the costly and burdensome chore of leaving every winter could be avoided, and mining would have the base to expand on a year-round basis. As the winter of 1873-74 settled on the San Juans, a new era dawned. Never again would these valleys and mountains be uninhabited; miners moving into them would find camps, mining districts, and mining laws awaiting them. The San Juans had many miles yet to go, but even the most pessimistic would have to concur that they had come a far distance since the Baker excitement in 1860-61.

[1] *Rocky Mountain News*, October 29, November 1, 9, 10, 21, December 4, 7, 10 18, 1860 and January 10, 15, 1861. Frank Hall, *History of the State of Colorado* (Chicago: Blakely Printing Company, 1890), volume 2, page 192. For a general overview of the region see Paul O'Rourke, *Frontier in Transition* (Denver: Bureau of Land Management, 1980).

[2] *Rocky Mountain News*, January 21, 1861. See also issues January 17, 19, 22, and 23. Edward Edwards to unidentified friend, December 1, 1860, Edward H. Edwards Letters, Beinecke Library, Yale. Wolfe Londoner, "Dictation," Bancroft Library. Harry Murphy, "Account Trip into Ouray Area, 1860," Colorado Historical Society.

[3] *Rocky Mountain News*, January 28, February 7, 9, 20, 22, March 14, and April 17, 1861. Virginia McConnell, "Captain Baker and the San Juan Humbug," *Colorado Magazine* (Winter 1971) traces Baker's career.

[4] Hall, *History of Colorado*, pages 193-195. *Rocky Mountain News*, January 17, February 9, and May 12, 1861, and (weekly), June 12, 1861. Alice P. Hill, in her *Tales of the Colorado Pioneers*, has several San Juan stories.

[5] Randolph Marcy to W. Nichols, August 29, 1867. Letters received New Mexico Superintendency, copy Center of Southwest Studies, Fort Lewis College. William Gilpin, *The Central Gold Region* (Philadelphia: Sower, Barnes & Company, 1860), page 140. J.P.Whitney, *Colorado* (London: Cassell, Petter and Galpin, 1867), page 45. For the pre-1860 years, (including Escalante quotes), see Herbert Bolton, *Pageant in the Wilderness* (Salt Lake City: Utah Historical Society, 1950), pages 26 and 140. J.N. Macomb, *Report of the Exploring Expedition* (Washington: Government Printing Office, 1876), page 82. *Silver World* (Lake City), September 18, 1875, *Ouray Times*, September 4, 1878, and *Engi-*

neering and Mining Journal, September 15, 1877, page 207, discuss finding evidence of previous people and mining. The *Santa Fe Weekly Gazette*, 1856-59, displayed very little interest in mining.

⁶ William Arny to Eli Parker, July 19, 1870, and journal entries May 19, 22, 25, 26, and June 11, 17, 22, 1870, letters received by the Office of Indian Affairs, microfilm copies Center of Southwest Studies. The April 20, 1870 *News* has an account of Indian Bureau activities.

⁷ *Rocky Mountain News*, May 18, 1872. See also issues April 21, 23, and May 12. Donald Brown, "Experiences in Southern Colorado," *Colorado Magazine* (January 1954), page 78. *Boulder County News*, April 26 and May 3, 1872. Hall, *History of Colorado*, pages 203-204. *Colorado Transcript*, April 24, 1872. For the Little Giant, see the *News*, April 21, September 4, 1872, February 18, September 28, November 16, 1873, April 3, and August 26, 1874. *Denver Tribune*, July 6 and August 21, 1872; *Colorado Chieftain*, April 6, 1872; *Engineering and Mining Journal*, February 11, page 44 and October 28, 1873, page 284.

⁸ *Annual Report of the Commissioner of Indian Affairs for the Year 1872* (Washington: Government Printing Office, 1872), pages 123-125 and 289-291. *Rocky Mountain News* (Weekly), August 21 and September 11, 1872. The Los Pinos Agency was thirty miles southeast of present Gunnison.

⁹ "Minutes of the Council," *Annual Report of the Commissioner of Indian Affairs for the Year 1873* (Washington: Government Printing Office, 1874), pages 106-107. "Narrative of Proceedings," pages 93-95; see also pages 16 and 83-85. *Rocky Mountain News*, March 9, 15, 30, May 9, 27, and July 12, 1873.

¹⁰ Ouray to Brunot, September 13, 1873, *Annual Report* (1873), page 86. See also Brunot Treaty, page 87 and "Narrative," page 95.

¹¹ *Mining and Scientific Press*, November 15, 1873, page 313. See also issues of October 11, November 22 and December 13, 1873, *Engineering and Mining Journal*, December 2, 1873, page 264. Chicago papers were quoted in the *Rocky Mountain News*, October 10 and December 24, 1873. *Utah Mining Gazette*, January 24, 1874.

¹² *Summering in Colorado* (Denver: Richards & Company, 1874), pages 141-142. For the San Juans in 1873, see *Colorado Transcript*, May 28, and July 2, 1873. Thomas Cree report, Annual Report (1873), pages 95-96. *Rocky Mountain News*, July 16, August 13, November 30, December 31, 1873, and May 2, 1874.

With deep snow blanketing the San Juans, the year 1874 took its bow in this, the first winter of permanent settlement there. For the miners huddled in Baker's Park, the days passed slowly in working their claims, constructing log buildings, dreaming of future wealth, and speculating how much snowfall might be "normal." These hardy fellows gained instant stature as "old-timers," experts ready to advise those who would arrive next spring. Less pioneering spirits in more settled regions predicted a rush as soon as the snow melted to passable amounts and

Ho, for the San Juans

the trails were opened. In his annual message Samuel Elbert, territorial governor, took note of the San Juans, praising their "marvelous promise" and, with a dash of political license, their rapid settlement and development. A local promoter identified only as "Southwest" gushed even more exuberantly, prophesying that the San Juans would be "the great excitement of the day," luring people from the east to the west coast. The *Rocky Mountain News*, March 4, brewed further intoxication by describing what the editor called the "mother vein," outcroppings of which has been traced nearly four miles. Tests on the ore had been highly satisfactory, almost "startling." Such stories precipitated mining stampedes. After all, it was a "known" axiom that silver mines get richer the deeper they are worked. Further inciting optimism was the maxim that "a good silver mine is above timber line;" high elevations were abundant around Baker's Park.

But no stampede engulfed the mountains that spring. Certainly interest increased, and there were definitely more people, but no genuine mining rush such as Colorado had experienced several times. Isolation and a questionable reputation served as dampers; unfavorable reports also hurt. One disgruntled returnee wrote that the men had staked every colored rock the previous year and, so far as he could learn, the country was all taken up. Prospecting, claiming, and reporting of high assays, rather than development and profitable mining, took up most of the 1874 season. One of those incurable optimists who seemed endemic to the San Juans concluded a long article in November by saying, "The mines are a success, and in a few years this will be, not one of the richest, but *the* richest and largest mining camp in the world."[1]

The remaining years of the 1870s fell short of his prognostication, years when the San Juans apparently were on the verge of becoming the most lucrative Colorado mining district, only to have some other area slip in and steal the headlines, glamour, and investors' money. Their high promise was restrained by poor transportation and poverty. Even so, in the national depression following the crash of 1873, a correspondent to the *New York Times* wrote, May 29, 1876, that it was healthful to have mining excitements: "our people in these dull times seem to be looking for wonders and feed on excitement."

Prospectors and miners trudged in, searching for their personal bonanzas, bringing with them an age-old expectation and a burro load of equipment. This should include, advised one commentator, three double blankets, a pair of hobnailed heavy boots, overalls, a blouse, a soft hat, coffee pot, bread pan, frying pan, tin plates, iron coffee mill, tin cup, table setting, and one substantial suit, apparently to wear on trips to town. Add to this, food (including dried fruits to balance the otherwise "heavy diet" of beans and grease) and mining equipment, and the San Juaner was off to the hills. The same writer recommended that $300 to $500 cash be taken along,[2] a wise precaution generally ignored. No one thought he would need money when it would be only a matter of weeks before silver and gold flowed into his pockets.

Drifting in and out with the seasons, as prospectors and tramp miners are wont to do (including Winfield Stratton, who had no luck here, on his way to Cripple Creek millions), they carved the main outlines of the San Juan mining region and at least scratched the surface by decade's end. The seventies spawned one little intradistrict rush after another, each promising more than was delivered. On the eastern rim, the Lake City area made its bid to become number one, its only serious rivals being Baker's Park (Silverton) and, in the San Juans' far southeastern corner, the Summitville mines. These three garnered most of the attention in the midseventies before Ouray grabbed a few headlines, a bit later because of its more isolated location. The San Miguel (Telluride) District also warranted some attention because of its placer gold deposits, but was even more isolated than neighboring Ouray.

Finally, as the decade closed, the Dolores River mines showed potential of their own under the banner of their leading camp, Rico. From these primary districts San Juaners spread out to nearby, but less well-known, discoveries. Near Lake City were Burrows Park, Rose's Cabin, and the mines along Henson Creek. Both Lake City and Silverton liked to claim the Engineer Mountain mines, the former having the momentary advantage. The latter was not hurt, though, because around Silverton were numerous mines, radiating like spokes from Baker's Park as far north as Animas Forks and south to the Needle Mountains. Southwest of Ouray was the Mt. Sneffels district and to the southeast Poughkeepsie Gulch, Red Mountain, and Mineral Point (Silverton and Ouray both claimed to be the trading center for this district). Snuggled between Rico and San Miguel, Ophir attracted fleeting attention, as did the mines farther south in the La Plata Canyon area. Just over this canyon's eastern rim the Junction Creek mines were opened, providing further proof that almost everywhere one prospected minerals could be found.

Whether traversing the canyons, wading the rivers, climbing the mountains, or crossing the alpine meadows, the San Juaners were relentless. If not this season, then next, the mineral secrets would be unlocked. Unfortunately, more than sheer enthusiasm and effort were required to turn potential into reality. Three basic ingredients have always been essential for creating a major mining region: profitable mineralization, capital, and economical and reliable transportation. By 1879 valuable mineral deposits had been uncovered, making this point academic; however, by themselves, these were not sufficient—there must be ore reserves to warrant continued mining. Capital—to develop claims, construct smelters, underwrite mining expenses and maintain settlement—had to be located. The third point was pivotal; without economical, year-round transportation the cost of living would remain high, only the smaller deposits of high-grade ore could be mined profitably, and investors would tend to shy away. Few mining districts reached full potential until this last question was resolved. These three factors interlocked, like the silver and gangue in the veins; the waste would have to be discarded and the rest refined in order for the whole to reach its potential and maintain its profitability.

In trying to solicit financial support, the San Juaners found themselves in an uphill struggle. In the first place, this was not the "poor man's diggings," that elusive spot which surfaced with the 1849 gold rush to California, where, by hard work and with simple equipment, a man could make his fortune. Hardrock mining cost money—one wag suggested that it took a mine to operate a mine. These pioneers were in trouble from the start. They had been warned that it was better to stay home unless they brought with them the means to live through a mining season. Few possessed so little faith that they could not anticipate uncovering a paying claim immediately, and fewer still had adequate finances to develop what they did find. So off they went.

In that kind of situation, once a claim was staked, the easiest thing to do was find someone else to buy it. This was not bad logic, because, at this point, a profit could be turned without the additional financial risk of developing a nonpaying property. Executing the idea proved more difficult. Investors did not flock to the San Juans, in spite of repeated pleas such as this in the *Silver World* (Lake City), August 26, 1876, "Come! Capitalists! Come! We are richer than you, but our money is locked in banks that never fail."

The midseventies' depression delayed interest in the San Juans. Then, just when some progress was being made, Leadville bounced into prominence in 1878, completely dominating the Colorado scene. At the same time the Black Hills and the Comstock diverted national attention. San Juaners tried desperately to turn those bonanzas to their own advantage by favorably comparing their mines to the ones at Leadville or the Comstock. Another device was to contrast the situation; for example, although the Black Hills had gained more attention, the San Juans had not had such a rush of disappointed miners "*out of the mountains.*"

Nor did the San Juans' isolation spur sales; it was easier to go to Leadville or Deadwood than to Silverton. The lack of home-controlled money for development meant that what were "mines" in the eyes of their owners became mere wildcat prospect holes with a chancy future in the opinion of investors. A favorite nineteenth-century expression called mining a lottery with few prizes and many blanks. Mine owners used myriad words and countless hours and local newspapers much ink, to try to dispel this concept.

A few investors showed interest, usually sending an agent to investigate and report. San Francisco mining engineer James D. Hague was employed by Tiburcio Parrott to find out why the latter's La Plata Canyon investment had not paid a profit. Hague's careful investigation spelled the end of Parrott's interest and investment and led to the decline of that little district, which had subsisted on California money, not gold from the river and the mountains.

The next year, 1877, another mining engineer, Joshua Clayton, visited La Plata Canyon, San Miguel, and Ouray on behalf of a client; his cryptic diary entries failed to disclose his opinions. Nevada senator John P. Jones, a noted mining investor, made his own investigation in August 1879, crossing from Lake City to his destination at Rico. He loved the scenery, was appalled by the primitive trails and the isolation ("might as well be in Africa"), and when sending his wife his agenda, concluded with "then thank the Lord we start for home."

Jones's investment, only in small properties, had much less impact than the appearance of Colorado's silver millionaire, Horace Tabor, with his Leadville-based fortune. Tabor, synonymous with success in Colorado mining, purchased mines at the head of Poughkeepsie Gulch and in the summer of 1879 toured the area, being accorded as nearly a triumphal reception as the San Juans could give. Tabor did nothing to dampen the enthusiasm when he described his property as "worth nearly or about the same now, I suppose, as my interests in Leadville" (worth several million). William Weston expressed the local hope: "Where the big fish go, the small fry all follow."[3]

Unfortunately, not enough of either materialized and some who did proved to be too cagey with their money. Ernest Ingersoll recounted an amusing story of the investor versus the eager miners. Camped in Baker's Park near Howardsville in the summer of 1874, Ingersoll was awakened by the "most diabolical shrieking and yelling," accompanied by rifle fire. Convinced an Indian attack was imminent, his camp prepared for the worst when it was suddenly confronted by dead silence. Next day the "attack" was revealed to have been only Howardsville's all-out welcome for an investor. The celebrity who caused all this commotion proved overcautious and reluctant. Local miners, unable to comprehend such reticence, grew more disgruntled by the day. Several weeks later, completely at rope's end, they staged a mock Indian attack, frightening their would-be financial savior into a posthaste departure.[4]

A few successful alliances evolved from this rocky courtship, generally in the form of companies which spread both the financial demands and the risks over a broader base. The Colorado Mining and Land Company, composed primarily of New York State investors, was organized in 1876 "strictly and legitimately for actual business," emphasizing the prevailing idea that mining was not a legitimate business. This group owned property in the Animas Forks-Mineral Point area. Like companies that would follow, it had a board of trustees and all the trappings of incorporation, including annual meetings and eastern directors. So did the Ouray Discovery and Mining Company, which controlled claims from Ouray to Baker's Park, and in its cheery pamphlet promised that work would start without needless delay. Mines, mill sites, and town sites could be obtained almost for a song, proclaimed the Lake City Mining and Smelting Company, which planned to acquire them once the money started rolling in from gullible easterners. These bargains would then be worked by its "conservative, honest and efficient" management. The plans proved premature when investors failed to be duped into buying stock.

To overcome such reluctance, the San Juans needed to be advertised and its mines and resources promoted, along with whatever other advantages over rivals could be conjured. The efforts of the seventies were similar to those tried elsewhere in Colorado, successes frequently proving cumulative rather than individual. E.M. Hamilton headed the list of promotional pamphlets with his *The San Juan Mines* in 1874, which included brief comments on mines, excerpts of local history, descriptions of natural wonders, and suggestions on how to reach districts. Hamilton, unlike some later writers, had visited the San Juans and based his work, at least partly, on personal observation. His purpose was to show that this was a bona fide mining region, to give an idea of the country, to illustrate the need for capital, and, finally, to "invite" capital. The Kansas Pacific Railroad published two pamphlets in 1876 and 1878 to advance itself as the gateway railroad. Similar to Hamilton's in intent, its literature included such practical information as how to file a claim, mining laws, and "how to make money mining." The *New York Times* had three articles, each two columns long, in mid-decade, giving warm praise and detailed descriptions. By the end of the seventies, the San Juans were also receiving mention, from one page to several short chapters, in publications like Pangborn's *Rocky Mountain Tourist* and Frank Fossett's varied treatises on Colorado. All together, these boosted and exalted as much as facts or imagination allowed, but they were merely drops in the bucketfuls of mining propaganda flooding the country.

These were a beginning but, by themselves, not enough. Maps showing routes into mining districts and mines (always the rich ones) were widely circulated. Speakers went forth panegyrizing San Juan virtues. One Ourayite spoke in New York City in December 1879, extolling the valuable mines, vast amounts of ore, and the immense water power reserves—nothing superior to them anywhere, in his estimation. Mining agents and owners advertised "bargains" in Denver newspapers and other nearby ones, and rich ore samples were displayed as early as the Territorial Fair in 1874. Local newspapers took it upon themselves to laud and boast vigorously and even reproach gently. The editor of the *Ouray Times*, August 11, 1877, scolded his readers for their lack of enterprise in collecting ore for an exhibition. He could not understand it, when so little effort would provide such a rare opportunity to excite men of capital.

Most San Juaners gamely promoted their region, as numerous letters to Denver, Boulder, Central City, and other newspapers testified. Some let their enthusiasm completely overcome reason; for others, disappointment clouded their impressions, but many gave a factual account about what they saw and did. One of the earliest, W.B. Dickinson, a man of frank and intelligent manners according to the *Rocky Mountain News*, corresponded with that paper in 1874. Far more influential was William Weston, one-time agent and writer for the Kansas Pacific Railroad, who moved into the San Juans in 1877 as a miner and assayer. Unable to still his pen, he wrote long articles to *The Engineering and Mining Journal* (New

The Highland Mary Mine in Cunningham Gulch was one of the earliest. The woman standing by the unfinished boarding house was a real pioneer of 1874-75.

York), as well as to Colorado papers. Weston, though suffering the misfortune of selling one of the really great mines (Camp Bird) before it came into bonanza, never lost his enthusiasm and, as late as 1910, was still reminiscing in print.[5]

The nature of the appeals was in the anticipation of wealth soon to be mined. Stressing mineral potential, this one-chord refrain, varying only in key signature, might have sounded monotonous, but it was necessary to make the San Juans at least a familiar, if not quite a household, phrase. Then and only then would investors come knocking, rather than having to be dragged in and convinced of local potential.

While that song was being sung, sharp debate continued, at a lower volume, between those shouting "Ho, for the San Juans!" and those crying "humbug." Part of this localized tempest reflected Baker's heritage, part of it genuine concern about misrepresentation, seasoned with a dash of jealousy. The *Mining and Scientific Press* switched from support to opposition in 1874, showing disenchantment with slow development and reports of mining troubles—then it lost interest completely. Typical of the humbug theme was the attitude of the *Boulder County News*, June 26, 1874, which quoted a disgruntled prospector's letter

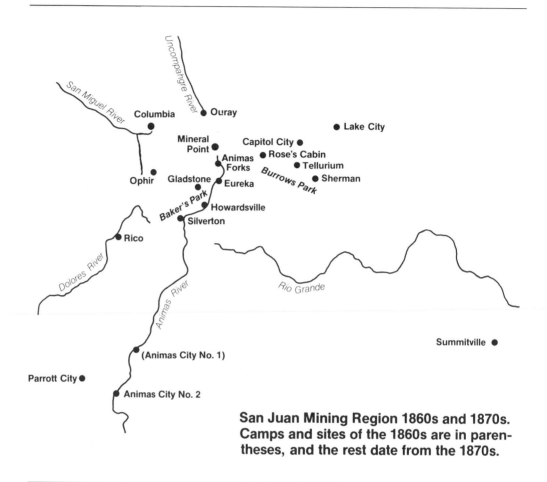

San Juan Mining Region 1860s and 1870s. Camps and sites of the 1860s are in parentheses, and the rest date from the 1870s.

advising no one to buy property or go there, when there was richer and more easily obtainable ore near Boulder. Supporters fired back, praising present progress and future prospects. An argument which neither side could win raged on, generating more heat than information. San Juaners probably paid little attention to it, caught up as they were in their own fantasies. For example, 25-year-old Alfred Camp, riding through the mountains, saw not rocks and trees: "As I jogged along on horseback in returning to camp, I thought of the great changes destined for this region—soon a large population would gather in this gulch, and smoke, possibly from hundreds of smelting works, would be seen ascending into the air."[6]

Already the changes Camp foresaw were occurring, perhaps not so rapidly as he and others would have wished. Though plagued by financial shortages, progress was being made in the all-important transportation realm. Without the improvements of trails and the construction of wagon roads, the San Juans would continue to stagnate. Until materials, equipment, and ore could be shipped reasonably, and until connections with the railroads or trading centers, such as Denver, took days instead of weeks, the San Juaner would find his mining difficult and his cost of living lofty, even for a mining district.

The main pioneering trail from Loma-Del Norte to Silverton, via Stoney Pass and Cunningham Gulch, continued to be a source of wonderment and agony to those who traversed it. William H. Jackson, noted western photographer, struggled up the steep slopes on a September afternoon in 1874. "What can possess those people we pass to go into that place this time of year?" he wondered. "We passed burro trains, and wagons. Smashed up wagons there . . ." A year later Camp described it as almost impassable, as hard to get an empty wagon out as a loaded one in: "This experience will hardly be forgotten as for several hours we scrambled and slid as best we could to get down—fortunately without breaking our necks." Even a local booster like W.B. Dickinson was forced to admit, "One begins to curse the roads immediately." The trout fishing along the Rio Grande, which paralleled part of the route, furnished him some consolation. If this main artery was bad, the intradistrict ones were far worse. One pioneer testified that on the Mineral Point to Ouray Trail in 1876 a man risked his life; it was not safe to ride a horse over, and when slippery was even unsafe to walk on.[7]

These appalling conditions could and would not be tolerated; an increasing hue and cry berated the miserable routes. Neither the territory nor the brand new counties possessed enough resources to underwrite improvement and expansion; fortunately, private enterprise stepped into the void in the form of toll roads. The most famous of their builders was Otto Mears, the Saguache merchant already instrumental in inducing the Utes to sign the Brunot Treaty. Ever alert to opportunity, he began building toll roads in 1874, starting with one from Saguache to Lake City, then branching from there up Henson Creek over Engineer Mountain to Animas Forks. In 1877, he opened a road that swept in a northwestward arch from his Lake City route and eventually reached Ouray. While this human dynamo started his San Juan transportation empire, others were working on shorter, only slightly less important, routes.

Thanks to everyday use, the trail hugging the canyon bottom from Animas Forks down into Baker's Park became a road. Without anyone to oversee it or provide maintenance, it proved a continuous source of worry and discussion. Some enterprising persons hacked out a road from Lake City to Burrows Park, then over to Animas Forks via Cinnamon Pass, which the miners helped maintain for their own interests. Silverton fumed over its reliance on Lake City and Del Norte and was gladdened in 1877 when a trail was pushed down the Animas River. While providing another outlet, unfortunately it did not go anywhere very significant at that moment. Every camp, out of a sense of self-preservation, wished to have a network of roads radiating from it. As a result, primitive trails and roads crisscrossed the older districts by the end of the decade. Rico, the newest one, bemoaned its lack of decent connections in 1879, yet within months the local editor pointed to wagon roads from the north and south and trails over the mountains. So great was the improvement of the San Juan transportation network that Rico, unlike its earlier neighboring districts, did not have to wait an entire season for at least passable roads.

Even with the engineering feats and labor that accomplished construction, criticism continued. Maintenance now became the overriding issue. Winter snows and summer rains, heavy traffic, and poor supervision wreaked havoc on road surfaces, a fact obvious to all toll payers as they were jostled and jolted about. Even Otto Mears, who had crews maintaining his roads, was subjected to criticism. Users complained about every imaginable problem, resulting in newspaper diatribes, protest meetings, and even individuals taking the situation in hand and providing repairs. The complaints were not all on one side; toll road owners found themselves up against a devious clientele, who concocted all kinds of schemes to avoid payment. Toll roads, nonetheless, answered a crying need in the seventies, though local appreciation was not guaranteed. Applause greeted the transportation arteries but not the toll costs. Eventually, public ownership was the answer; the peak of toll road building had been reached in the late 1870s and early 1880s.

While road discussions and construction were attracting the most attention, visionary San Juaners turned to the railroad as the ultimate solution. As early as January 1874, a heady optimist forecast railroad connections before a "great lapse of time," and a little later the same year Ernest Ingersoll correctly predicted that when the railroad came it would come up the Animas Valley. The seventies generated only plans, such as the route proposed from Baker's Park to the coal deposits of the lower Animas Valley, and wishful thinking about the blessings the iron wonder of the age would bestow. Locals heartily agreed with the *Dolores News* editor, November 22, 1879, that all that remained to be done was to push the railroads to the mines and the San Juan would come of age.[8] Be that as it may, the less romantic freight wagon, sturdy mule, and burro kept the San Juans supplied while dreamers dreamed.

Meanwhile, mining spread throughout the region, engendering the growing pains typical of any new district. Placer operations, except on the San Miguel River, declined and were replaced by quartz mining. This in itself explained the demand for more money. Now the miner needed to purchase timbers, rails, and mining cars, blasting powder, drills, and other equipment to construct a hoist, dig a shaft, or drive a tunnel. He could not rely on a pick, a shovel, a pan, and a strong back. New skills, usually acquired by trial and error, were also necessary. Importing experienced miners was faster but more costly. Ventilation and underground water never bothered the placer miner; they hounded his hardrock compatriot.

Two districts quickly sprang to the fore and maintained that lead throughout the seventies: Lake City and Summitville. Located on the San Juans' eastern rim, both were more accessible in all respects, and they thrived on this advantage.

Lake City came first, spurred by rich surface deposits, Mears's toll road (much better than Stoney Pass), and its own claim to being the gateway to the San Juans. Attracting attention from the start, it moved into its first mining boom in 1874, securing both outside investors and reduction works before neighboring districts had time to bat an envious eye. Its mines became better known and developed, primarily the Hotchkiss (Golden Fleece) and Ute and Ule. Lake City, with its clear advantages, parlayed them masterfully for several years.

But both mines provided only short-lived bonanzas. The Hotchkiss (a gold mine in a mostly silver region) hit pockety ore, a vein inconsistent in value, and soon declined. As early as 1877 it was being described as a famous mine where work would soon resume, a sure sign of adversity. The Ute and Ule (a pair of mines eventually worked as one) were to have a much more distinguished career but not in the 1870s, when inexperienced management, high developmental costs, and inconsistent ore deposits put them into periodic trouble. Other so-called mines—holes blasted a few feet into a hillside or shafts just barely below the surface—proved to be rich only at the surface or valueless, and soon were abandoned or put on the market.

Here, as elsewhere, the majority of the original locators did not have money for long-range development, nor did the transportation advantage benefit the whole district equally. The farther the mine was situated from Lake City proper, the higher the freighting costs and generally worse the road, until, when reaching Burrows Park on the western edge, the costs and isolation weighed heavily. Flurries of activity that brought brief distinction to Rose's Cabin and similar spots did not insure productive mines.

The contagious excitement of the opening months (122 lode and 8 placer claims filed within weeks after the organization of one mining district) waned, leaving Lake City's mining in the doldrums, barely maintaining its leadership by 1879. The *Rocky Mountain News*, June 22, readily blamed too little money, weak faith, and apathy. Lake City's troubles, however, rested with its mines, whose ore only tantalized but did not sustain consistent production.

Silver generated Lake City's appeal; gold inspired Summitville's. The 1874 fever resembled earlier Colorado gold rushes, which were sparked by rumors and some freely displayed gold nuggets. Placer deposits panned out quickly, which did not worry one local enthusiast who described the neighboring mountain as "one vast gold deposit." Quartz mining

The North Star Mine was typical of the early small operations. Mules and burros were the supply carriers to these isolated spots. A burro could manage 200 or more pounds and a mule 250 up grade, 350 down grade.

Lumbering was a significant support industry to mining and settlement in general. This portable mill was set up west of Telluride. Lumbering denuded much of the region, as the need for wood was great.

replaced placer, and the appearance of stamp mills in 1875 left Summitville with few reduction worries, its treasure now recoverable by proven methods. Soon hailed as southern Colorado's richest gold district, it was doubly blessed by secondary enrichment, or the natural process of leaching out impurities near the surface, leaving behind valuable minerals in easily reduced form. This zone produced most of the gold, over $700,000 worth, during these years.

In contrast to these two districts' successes were the problems encountered in trying to establish placer operations along the San Miguel River. Prospectors are known to have reached the valley in 1875; a few even wintered there. The next summer, larger groups worked placer claims, inaugurating permanent mining. Acres of river bank and bottom were staked some thirty miles along the San Miguel by 1879, and encouraging progress reports were carried over the mountains. From pan and shovel, the operations progressed through sluice boxes to more refined methods until large-scale hydraulic operations, similar to California's, were being planned by companies which controlled vast acreages.

Still the gold proved elusive. Large rocks interfered, as did unreliable water reserves, without which work stopped. Most significant, the area was one of the San Juans' most isolated, the trails into it being long and poor even by standards of the day and time, pushing up mining and living costs and slowing all shipments. The mountainous elevations gave only a few working months between spring thaw and winter freeze. The companies manifested more hope than financial resources, another detrimental element. By decade's end they found themselves as frustrated as the miner with his small claim, which he desperately tried to work during the warm summer months. Somewhat ignored in all the excitement over placers were the mines located in the surrounding mountains, but their discoverers were no better off than their co-workers down on the river bank.[9]

From the three foregoing examples, the problems of mining in the San Juans can be clearly discerned. One major factor perhaps was not so evident. Working with gold had not caused major reduction problems (e.g., Summitville). Silver, however, had generated most of the San Juan enthusiasm, and reduction of this metal was troublesome to miners, who did not understand how to reduce it economically and profitably from ore to bullion.

San Juan ores, diverse in their mineral complexity, stubbornly refused to yield to simple reduction methods, a characteristic that earned for them the terms "rebellious" or refractory. Two avenues were open for finding the best method. The ore could be shipped out for testing by various processes to find the one that would save the largest percentage of its assayed worth. Or a process could be brought in which it could only be hoped would be successful, or at least modified to work. Because the latter alternative was faster, though not necessarily wiser, San Juaners chose it; they were not about to wait any longer than necessary in such a crucial matter.

The arrastras and other simple devices of earlier days would not suffice, nor would that ultimate expedient, shipping only the richest ore out to be worked. One small shipment did, however, go all the way to smelters in Swansea, Wales. Smelting, the reduction of metals by heat in a furnace, seemed to be the best available solution. Silverton, closely pursued by Lake City, struggled to become a smelting center. Silverton's first smelter (Greene's), apparently built in 1874 or 1875 (the date is immaterial since it was the earliest in any case), was a lead-based smelting operation which concentrated the ore, removing the gangue or worthless matter. The matte, as it was called, was then shipped elsewhere to separate completely the silver, lead, and small quantities of gold and copper. The Greene works led an erratic existence, frequently interrupted by shutdowns. James Hague visited it during one of those idle periods in September 1876. Though impressed with the buildings, he commented that the ore supply appeared to be pretty short. Each season seemed to necessitate some overhauling, repairs, or modification until finally, in 1879, the firm collapsed. It was sold at a trustees' sale and became part of the San Juan and New York Smelting Company.

Greene and his partners paid the price for being pioneers. They found, for instance, that many local ores were unsuited for their process simply because there was not enough silver-lead ore available for flux. In addition, the high costs of coal and coke (often of poor quality despite the price) and freighting meant that operational expenses ran high and success low. Their successors hoped to overcome at least the ore shortage by combining mines and smelter into one company.

This same idea had been adopted at Lake City by the Crooke brothers, who ran the most successful smelter in the 1870s. The Crookes, who owned a smelter in New York City, had solid financial backing and experience. Contemporary mining reporter Frank Fossett credits them with being the first eastern capitalists to show real interest in the San Juans (they invested in the Lake City and Summitville districts). Their interest had been whetted by favorable reports from the area. The prospects were confirmed by a personal visit from John Crooke and others in 1876. They built a concentrator that year and shipped its products to New York to be refined. This process was improved in 1877 with the addition of reverberatory furnaces to separate silver from the lead bullion without shipping it elsewhere. Broadening the company's involvement, they purchased mines, including the Ute and Ule. Because of their previous experience, sounder financial structure, and conservative approach, plus Lake City's better transportation, the Crookes were able to succeed (Fossett estimated $112,121 bullion produced in 1879) where Greene had failed.

The Crookes had several rivals. Scattered throughout the San Juans were other smelters: the Ocean Wave, which based its hope on the mine of the same name; the Capitol City, built in 1877 at the camp above Lake City but crippled by ore shortages and poor management for several years; and the Van Gieson Lixiviation Works at Lake City. Lixiviation, a method based upon roasting the ore, then introducing various salts to leach or wash out the base metals and precipitate silver, gained a measure of popularity with the San Juaners. The San Juan Lixiviation Company laid out the town of Gladstone, north of Silverton, in 1877, and in 1878 opened its plant; the process, however, had not proven successful by the close of the seventies. Also providing competition for Greene were concentration works at Animas Forks and in Cunningham Gulch, on the route to Stoney Pass. Fleeting references to additional works appeared in the papers, indicating that the San Juans were well "smeltered." Too well, actually, since all of them were trying to work a rather limited supply of ore. The Rough and Ready smelting works at Silverton proved neither of its adjectives correct and went under because of "financial embarrassment" in less than a year. So did those idealistic souls who backed the "lighting amalgamation" process based upon stamps and amalgamation; they may have saved some gold, but they lost their shirts on silver.

Enough smelters? "No!" shouted Ophir, Rico, Sneffels, and Ouray. "We want our own! Our mines are idle because no local works exist." Ouray, with aspirations to become the leading mining center, was especially aggrieved, since smelters were felt to be essential to any major district. When several different processes failed, the *Solid Muldoon*'s editor, Dave Day, was prompted to complain on October 10, 1879, that the Ouray camp, while capable of producing hundreds of tons per day, had no home market for a pound of ore anywhere near its real value.

Day put his finger on a continuing vexation, the inability of local smelters to pay for the ore at a price close to its assayed value. The smelters discounted for expected loss of mineral in processing, transportation expenses, and their own profit. This left the miner holding a bag of depreciated ore and unanswered questions. Hague reported that the Greene smelter claimed to pay 70 to 80 percent of value; not so, Silvertonians told him, it was generally 50 to 60 percent. One obvious reason for the ore shortage was the fact that the miner could afford to sell only his high-grade ore, leaving most of the ore sitting on the dump.

For San Juan smelting this was an experimental decade; no answers came forth to solve

Summitville was a gold area and the Little Annie mine shown here one of the better ones. One of the important early mining areas, Summitville was finished as a major district in a decade.

many of the pressing questions. This experimentation had been costly in time, in investors' financial losses, in dashed hopes, and in gold and silver washed down a nearby creek or dumped on the tailings pile. Until the mine owner could be assured a fair price, mining was going to stagnate.[10] A promising development, single ownership of mines and smelters, seemed attractive, if the mine could produce enough to absorb the lion's share of a smelter's operation. They were all jammed together—smelting, capital, transportation, mining expenses—and the San Juaners awaited the break that would blast them free.

A quick scan of the 1870s shows much prospecting (most San Juan claims were filed on during this decade) and opening of mines. A mine developed to the depth of forty feet, or opened by a tunnel of like distance, was judged to be well along. Considering the meager resources of many of the owners, this probably exemplified the best that could be done without outside help.

As did the rest of Colorado, this region went through a tunneling craze. Driving a tunnel into a hillside looked to be so much easier, a cheap way to tap veins and drain a property. No need to rely on expensive, time-consuming shaft digging and installation of pumps and hoists. The idea hit the San Juans just as the seventies closed, particularly the circuit from Mineral Point to Silverton. It seemed to augur well for the San Juaners, who eagerly grabbed at any likely panacea.

Meanwhile, few large sales were transacted and the Little Giant, one of the earliest to change hands, became the aforementioned first major fizzle when, in 1874, work stopped on this previously proclaimed "fabulous mine." Law suits, which came to haunt San Juan mining, harassed this operation. Charges of mismanagement and a pinched-out vein surfaced, writing *finis* to the Little Giant's role as a major mine. Only bewildered stockholders were left to analyze the reasons.

Hydraulic mining was centered along the San Miguel River. No matter what year, it produced an environmental mess, both at the site and downstream—and little profit for investors here.

Fortunately for the San Juans' reputation, failures of the magnitude of the Little Giant were few, but wildcat speculation and fraud became rampant. In a situation where would-be experts raced everywhere, their pockets bulging with glittering samples and their sales pitches at the ready, it was not surprising that a con man or two managed to find a hearing. Local newspapers patrolled vigilantly, repeatedly cautioning readers about worthless claims, "miners' scalpers," or charlatans. William Weston and Dave Day warned residents and easterners of fraudulent companies operating around Ouray with fancy prospectuses and exorbitantly priced stock. Weston admonished prospective buyers to follow three rules: go see the mine, take along a man of mining experience you can trust, and select your own samples from the pay streak. Outspoken Day, in his *Solid Muldoon*, laid it on the line— those who got "sucked in" had "no one to blame but themselves."[11] This was only partly true. The naive easterner, convinced by the seeming authenticity of a company's presentation, might accept statements at face value. One particular company that aroused Day's ire was also exposed by the *Engineering and Mining Journal*; even so, not many people were aware of the actual situation. With no policing agency except public opinion, the field stayed wide open. The problem threatened to undermine all the efforts of honest San Juan promoters; it could not be ignored for long.

For the working miner, one season was much like the others—an arduous, dangerous life. Year-round work became more common, the mining boarding house now part of all the large, isolated mines. These could be a godsend for the night-caught traveler, who generally found warm hospitality. The cook, huddled in the kitchen, was nearly as important to a mine's reputation as its ore. Unemployed, William Rathmell hired on as a cook, hoping that one of the miners would leave and he could replace him. Since he knew nothing about cooking, his efforts first received grumbles, then sarcasm. One particularly bad day, after burning the bread and forgetting to cook the meat, he quit, leaving his boarders to ill-concealed relief. Obviously his efforts were not typical of the trade, or a culinary revolt would have flared throughout the San Juans. Faced with a hungry crew each morning and evening, only the creative, or perhaps the thick-skinned, survived. Pity the poor cook whose winter supplies had not all arrived before the storms settled in, or the one who misjudged the miners' appetites when ordering. Monotonous menus abetted eager complainers, not eaters.

Another change, this one in the overall employment pattern, became evident. Men were hired by a company instead of working for themselves or on a small two-to-five-man operation. The man who owned his own claim and mined alongside the few miners he could afford to hire predominated and retained a camaraderie between management and labor that was difficult to maintain in a larger operation. The San Juans still offered the possibility of working briefly for someone, gaining a stake, then moving out on one's own. As the years passed, this hope became more fleeting—districts were filling up and profitable claims were hard to find.

The typical miner was of American, English, or northern European stock, if the returns of the 1880 census may be projected backward. He was, according to those who knew him, generally young, poor, generous, and single. Patience, courage, and strength were traits he needed to succeed, one writer noted. Alfred Camp left a picture of his companions in 1875. One had mined in California, and also been a sailor; another was

a great tall lank individual. His clothes hanging about him in rags. A broad brimmed greasy hat covered a head of hair that had not seen a comb or brush for months—his beard matted and tangled like the mane of a lion.

Most observers did not try to distinguish between a prospector and a pioneer miner, and it mattered little, because some men readily shifted back and forth and the traditional prospector was vanishing, moving on to unexplored land.

Generally an individualist, the miner would organize for mutual protection by establishing a mining district, for example. However, these were less important now that the federal and state governments had mining laws and counties had been organized. Miners' meetings were held occasionally to handle such issues as claim jumping and registration, the former an explosive issue if it persisted. These were not years of unionism, although at least two local groups flourished for short periods in Lake City and Ouray-Mineral Point, apparently to maintain wages for experienced miners. [12]

Mining was firmly established in the San Juans in the seventies. Silver and gold ranked first and second in production, followed by those metals mined as by-products. Already there was growing interest in coal. Both Joshua Clayton and James Hague reported coal outcroppings in the La Platas and Animas City (or lower Animas River Valley), and the San Juan and New York Smelting Company was interested in coal veins in the same area.

In these first six years of hard rock production, over $2,200,000 had been mined, [13] compared to over $11,000,000 from Leadville in 1879 alone, or older Gilpin County's steady $2,000,000 per year average. For this period, as for later ones, no exact figures can be determined, largely because of the failure to keep accurate records and the tendency to overestimate local production. Hinsdale, Rio Grande, and San Juan, ranked in that order, were the three big counties, producing approximately 90 percent of the total. One intriguing question was whether production represented a surplus over opening and operating mining expenses. For the 1870s, if the estimated man-hours involved in mining and opening the region are compared to production, it is evident that the San Juans operated at a deficit. The investor found himself putting in, not taking out.

Mistakes had been made, lessons painfully learned. The *Engineering and Mining Journal*, December 20, 1879, pointed them out: mining gluttony, especially in building too many smelters; erection of costly machinery in an experimental manner; overestimation of claims; and the circulation of reports without the slightest foundation in fact.

A fascinating decade had slipped away, one of silver dreams. While the San Juans never attained the stature of "the promised Silverado" predicted for them back in 1874, the expectation persisted, now a little more soundly based on fact. This current of excitement carried on into the 1880s. Amid the mounting drifts of snow nestled around his cabin in that winter of 1879-80, many a San Juaner agreed with the analysis in the *Denver Tribune*, January 1, 1880:

> The future of the San Juan cannot be predicted. Its wealth is unbounded, its mines are almost innumerable as the leaves of the forest, its valleys as rich and fertile as any in the Union, its climate salubrious [this might have produced an argument], and its soil rich and productive.

1 *Rocky Mountain News*, November 1, 1874. See also issues January 4, 7, March 4, April 3, May 2, 9, September 18, and October 8, 1874. *Boulder County News*, June 26, 1874. *Weekly Arizona Miner* (Prescott), February 6 and 27, 1874.

2 *Silver-Seeking in the San Juan Mines* (Kansas City: Ramsey, Millett & Hudson, 1878), pages 18-19.

3 *Engineering and Mining Journal*, December 27, 1879, page 469. *Dolores News*, August 21, 1879 and December 31, 1881. *Denver Tribune*, April 17, 1879. *Rocky Mountain News*, July 15, 1879. John Jones to wife, August 3 and 6, 1879, John P. Jones Papers, Henry E. Huntington Library. Joshua Clayton Diary, July-September 1877, Bancroft Library. James D. Hague to Tiburcio Parrott, October 12, 1876, James D. Hague Collection, Henry E. Huntington Library.

4 Ernest Ingersoll, *Knocking Round the Rockies* (New York: Harper & Brothers, 1899), pages 122-124.

5 Material by Weston can be found in the *Engineering and Mining Journal*, particularly the late 1870s and early 1880s. Sources for the preceding paragraphs are: *The San Juan Mines* (Kansas City: Commerce Publishing House, 1876); *Silver-Seeking*; *Rocky Mountain News*, January 25, September 24, 1874, April 29, June 29, 1875, and December 13, 1879; E.M. Hamilton, *The San Juan Mines* (Chicago: C.E. Southard, 1874); *Colorado Mining and Land Company* (Buffalo: Baker, Jones & Company, 1876); *The Ouray Discovery and Mining Company* (Kansas City: Ramsey, Millett & Hudson, 1879); *Lake City Mining and Smelting Company, San Juan* (New Haven: Hoggson & Robinson, 1876), and *New York Times*, May 29 and November 3, 1876 and January 8, 1877.

6 Alfred Camp Journal, August 2, 1875, Colorado Historical Society. References to the debate are found in *Mining and Scientific Press*, February 21, page 123, March 14, page 173, and April 4, 1874, page 218. *Rocky Mountain News*, August 1 and 26, 1874, and October 8, 1875. *Engineering and Mining Journal*, February 21, 1874, page 122 and July 14, 1877, page 30. *Weekly Arizona Miner*, March 13, 1874.

7 LeRoy and Ann Hafen (eds.), *The Diaries of William Henry Jackson* (Glendale: Arthur H. Clark Company, 1959), pages 328-329. Camp Journal, August 3, 1875. *Rocky Mountain News*, November 1, 1874. *The Board of County Commissioners of the County of Ouray versus . . . San Juan: Defendant Brief* (Denver: Clark Printing Company, c 1908), page 49.

8 The section on transportation was taken from various issues of the *Rocky Mountain News*, *Ouray Times*, *Silver World*, and *Dolores News*.

9 For Lake City see *Silver World*, 1875-79; *Engineering and Mining Journal*, 1874-79; *Ouray Times*, August 24, 1878. For Summitville see Fossett, *Colorado* (1876), pages 423-434; *Rocky Mountain News*, November 27, 1875, June 1, 1876, July 22 and November 8, 1879; *Saguache Chronicle*, June 10 and August 5, 1876. For San Miguel see *Ouray Times*, September 15, 1877, August 31, and September 14, 1878; *Solid Muldoon*, October 24, 1879; *Silver World*, October 30, 1875 and August 12, 1876.

10 Material on San Juan smelting can be found in all the newspapers listed in previous footnotes. See also Parrott City Notebook, September 18, 1876, Hague Collection. Fossett, *Colorado* (1879), pages 516-517. *San Juan and New York Smelting* (New York: Jones Printing Company, 1880), pages 5-9.

[11] *Solid Muldoon*, October 17, 1879. *Engineering and Mining Journal*, November 15, page 363 and November 22, 1879, page 378. See also *Ouray Times*, August 17, 1878, *Silver World*, March 4, 1867. *Dolores News*, August 28, September 30, and December 6, 1879. *Rocky Mountain News* (weekly), April 17, 1878. *New York Times*, May 29, 1876.

[12] All the local papers have comments and stories on miners. See also I. Smith Letter, May 12, 1876, Colorado Historical Society. George Darley, *Pioneering in the San Juan* (Chicago: H. Revell Company, 1899), pages 16-19 and 69-70. *San Juan Guide* (Topeka: Atchison, Topeka & Santa Fe, 1877), page 1. Camp Journal, August 1-2. William Rathmell, "A Brief History of Ouray County," Western History Department, Denver Public Library. For an excellent overview see Ronald C. Brown, *Hard-Rock Miners: The Intermountain West* (College Station: Texas A&M Press, 1979).

[13] This estimate is based on county-by-county production totals in Charles Henderson's classic study, *Mining in Colorado* (Washington: Government Printing Office, 1926).

To the casual observer of the 1870s, it seemed that San Juan mining camps sprang up nearly as often as mines were discovered. New ones burst into being every year, fortified with youth's optimism and confidence, each convinced that ere long it would be *the* metropolis. At least twenty-nine received a baptismal notice in some newspaper. How many more might have been still-born, passing without notice and mourned only by their promoter-fathers will never be known. In the peak years of 1874-75, a fresh crop blossomed every spring, the less hardy ones

Climb the Mountains High

fading within a season. After those years the numbers leveled off, until by 1879 Rico alone was worthy of notice.

Geography and environment dictated, but did not always limit, the selection of camp sites. Natural parks or bowls were quickly snatched, as evidenced by Silverton, Columbia (Telluride), and Ouray. Mouths of canyons, serving as gateways to transportation routes or mining districts, proved equally popular. Howardsville and Parrott City were examples. Even sites at lower elevations, near farming land or mineral springs, were developed, such as Animas City (No. 2, located seventeen miles south of Baker's original settlement) and Wagon Wheel Gap. Lake City's location combined several of these features, situated as it was in a sheltered valley at the vortex of transportation outlets and serving as a portal to several mining districts. Eager town builders were not discouraged if their location possessed none of these natural advantages; all they required was that it be at the mines or on a road toward them. Thus camps were built where logic or geography said they never should appear (e.g., Mineral Point, Tellurium, and Summitville).

Visitors to the San Juans in the seventies often received a general impression that did not please dedicated San Juaners. Even mining engineers somewhat accustomed to roughing it were stunned. David Brunton went to Mineral Point, an "alleged city," in the summer of 1875, where rainfall pelted him and snow flurries were so frequent that a tent proved far from comfortable. Snow closed the trail before winter provisions arrived and Brunton, with considerable effort, tramped out to Lake City. He and a friend sat down to a disappointing restaurant meal, climaxed by dried apples for dessert. With great contempt his companion vented his feelings: "Damn a country where dried apples are a luxury." After four months in the wilds of the San Juan, Denver seemed immense—its hotels the height of luxury—to the returning Brunton, who had criticized both on his inward trip. To James Hague, after a restless night in a poor hotel, Silverton was a comfortless place of frame and log houses, and William Henry Jackson passed through Elbert in 1874 without even knowing it. Senator John Jones, although critical of the trails, isolation, and the primitiveness that reminded him of early California, was captivated by the beauty, writing to his wife in 1879 that the scenery "has been magnificent and grand beyond my powers of description, far surpassing anything I ever saw in the Sierras."[1] Jones was right; though aesthetic considerations played no part in the selection of town sites, some of the most strikingly beautiful mountain grandeur in Colorado ringed these pretentious little settlements.

Regardless of the setting, the communities would not have emerged had there not been a demand for them. The San Juan prospector and miner, like his counterpart elsewhere, was not a self-sufficient individual who could survive by his own efforts for very long. He needed supplies of all kinds. Not wishing to stop the feverish search for precious metals, he naturally looked to others to provide his necessities. Furthermore, he had the money, either borrowed or dug from the ground, to buy those goods. And so, on the miners' heels, came those who "mined the miner," and with them the camps emerged.

Why some camps prospered and others faded is easily explained. A well-chosen natural setting was the prime factor in Lake City's rise to early prominence as the leading San Juan community. Blessed by a location lower and more sheltered than all the other nearby camps, it was also nestled in a valley large enough for ranching and farming. Water power for smelters and other industries flowed past its doorstep, and its access to transportation has already been mentioned. Soon a network of roads fanned out, spokes from a business hub, tapping all nearby mining districts in the very heart of the San Juans. It also provided the best route to the outside, via Saguache. Lake City's founding fathers did not waste these attributes, parlaying them shrewdly to their advantage in the mid-1870s.

Silverton also occupied a hospitable natural setting that allowed for easy expansion, although it was not much better than neighboring Howardsville and Eureka, which were both located nearer the mines. Howardsville had the further advantage of being at the mouth of Cunningham Gulch, the end of the Stoney Pass road, and it was older by several months. The swing factor in the 1870s that tipped the scales in Silverton's favor was the aggressiveness of its founders and early merchants. When the earliest smeltermen came into the area they went to Howardsville to select a site, only to find land overpriced as the locals overplayed their hand, confidently believing the smelters would have to locate there. They were wrong. Silverton provided cheaper sites and nabbed the smelters. Sawmill operators were likewise enticed to locate nearby, and before long Howardsville and Eureka residents came to Silverton for lumber and ore processing. To overcome their isolation in the heart of the San Juans (one visitor remarked that "one feels as if he is eventually shut out from the rest of the world as there seems no way out"), ambitious Silvertonians launched a road-building program, which quickly brought results with a trail down the Animas Valley. Howardsville was designated the county seat, but Silverton was not dismayed; within a year it had electorally captured the honor from its rival. Within three years after its founding (1874), Silverton, by aggressiveness and farsightedness, had built a foundation from which to launch its drive to dominate the local urban scene. The keys were these: smelters and sawmills (vital industries in a mining district), the county seat, and direct transportation to the outside. As a result, the area's shopping patterns changed, focusing now on Silverton, which provided most of the services. Howardsville and Eureka never recovered, becoming satellite camps of their larger neighbor.[2] Their day had passed, almost before it got started.

Aggressiveness alone gave no guarantee of success. Few pioneer San Juaners did more to create a camp than did John Moss for Parrott City at the mouth of La Plata Canyon. Following his successful negotiations with the Utes for land, Moss turned to his California backers, the Parrott banking family, for more money, which was soon forthcoming. In 1874-75 Moss had a free hand to promote his camp, which he did, calling attention to his discoveries, isolated though they were. Parrott City, designated the county seat of La Plata County in 1876, acquired a post office and received such editorial praise as "beautifully located" and "lively," while its founder was honored by being elected to the first state House of Representatives. Beauty was in the beholder's eye in this case. William Henry Jackson, coming through in 1874, said Moss's camp consisted merely of a few small tents and some brush wikiups (frame construction came later) and put his finger on the reason for its existence—the Parrotts' financial support. When no profits materialized, the worried investors dispatched mining engineer James Hague to investigate; his 1876 report on the mining potential ended the Parrotts' interest. The congenial Moss, realizing his financial fountain had dried up, left his constituents high and dry, decamping from Denver during the legislative session to head for California. Parrott City stagnated, maintaining its hold on the county seat only because no rival challenged it. Its hopes as a mining and commercial center flickered out with the Parrotts' withdrawal.[3] Isolation, lack of nearby producing mines or placers, and the cutting of the economic umbilical cord doomed its future despite Moss's energetic promotion.

Some of these hamlets came into being without the advantage of nearby mining. Wagon Wheel Gap began as a station on the Del Norte-to-Silverton road and became, by the second half of the 1870s, the San Juans' first tourist attraction, famed for its hot springs and fishing. Nestled around the springs were bath houses, "plunge baths," cabins, and an "elegant little hotel." An "even more efficacious" cure was promised the patient willing to be buried in the thermal mud springs and left there until the "poultice draws out" the disease.

During the urban growth of these years clusters of mining camps developed, subservient to one trans-shipment or commercial center. The two most prominent groups were dominated by Lake City and Silverton. Following Silverton's surpassing of Howardsville and Eureka, all other camps within a reasonable traveling distance also came under its influence, including Gladstone, Animas Forks, Neigholdstown, and, for the moment, Mineral Point. The area encompassed roughly fifty square miles (the upper Animas River, Cement and Cunningham creeks), all of which drained into or through Baker's Park, where Silverton stood. Lake City dominated more than 100 square miles, primarily the drainages of Henson Creek and the Lake Fork of the Gunnison River. In the latter was Burrows Park, serviced by three tiny camps (Sherman, Argentine, and Tellurium), each dependent upon Lake City, twenty miles distant by mountain road. Sherman, and others mentioned, had their own limited business sphere or zone of influence, and occasionally an even smaller camp depended on them for supplies. One thing sure to arouse local animosity was an attempt by a dominant camp to tap another's tributary district, something Lake City tried to do in the seventies by expanding its toll road system. Predictably, its rivals, Silverton and Ouray, rose up in anger.

By the end of the 1870s, Ouray was beginning to challenge the older two and would represent more of a threat once it could acquire shorter connections to the outside. The rest of the San Juans proved somewhat a "hit or miss" proposition. The Summitville area depended on Del Norte, which otherwise was rapidly losing its once-important position as the entry to the San Juans. Saguache, with its new shorter road to Lake City, replaced Del Norte and remained the principal trans-shipping point. The La Plata district looked either to New Mexico or Del Norte for freight and Silverton for mail, a long haul regardless, and the western districts—Rico, Ophir, and San Miguel—vacillated between Silverton and Ouray.

In physical layout, the camps varied. Animas Forks was strung out along one main street; Ouray was blocked in a grid pattern, horseshoed by the bowl in which it resided. A company planned Eureka, which still was mostly on the drawing board, and Tellurium was scattered indiscriminately beside the main road past its site. Silverton and Lake City had enough room to develop along the traditional grid pattern so common to earlier frontier settlements. Natural growth predominated over a smattering of planning. Even the best of plans, however, could not compensate for a lack of expansion or the fact that some "unplanned" street became the business heart of the camp.

Into these communities poured a variety of San Juaners, no longer real pioneers, that stage having ended in 1873-75. They were the second wave of settlement and would not find conditions as rough and crude as those who immediately preceded them.

The camps themselves were dominated by the emerging business class, hard to identify individually because the principal sources, business records and newspapers, are skimpy or nonexistent for this period. The business district, the camp's heart, would be one of two types: the generalized one of the "little hamlet" or the more specialized one of the larger settlements. For instance, in the numerous smaller communities a general store, saloon,

Prospectors and miners coming into the San Juans lived in camps like this one in La Plata Canyon. They hauled with them all the supplies and equipment they needed for a season of mining and probably left before winter closed in with its cold and snow.

and possibly a blacksmith and hotel composed the core. There might be more than one of each, or several could be combined under a single roof, overseen by a budding entrepreneur. An assay office, boarding house, and/or restaurant added signs of progress and, unless a specialty store gained a foothold, rounded out the business district. The backbone was the general merchant, who carried a smattering of many products, from patent medicines to miners' supplies, but did not offer an overabundant selection to the shopper. Generally, he arrived first and his store often became the nucleus around which the camp grew.

Specialized businesses usually located in the larger settlements, making it more attractive to shop there, thereby increasing trade and encouraging more tradesmen. This cycle propelled these communities farther ahead of their less prosperous neighbors. In the 1870s four camps flourished: Lake City, Silverton, Ouray, and Rico. Some businessmen seemed to possess a sixth sense for detecting influences which would allow one community to grow and another to stagnate; these individuals gravitated toward the four camps mentioned. Rico was not even a year old in 1879 when its paper (something many San Juan communities never acquired) listed the following specialized businesses: tobacco and cigar store, barber shop, baker, meat market, livery stable, hardware store, stationery, and drug store.

The only town for which a very thorough examination of an emerging business district can be made is Lake City. The natural advantages of its site were quickly perceived in 1875, when its business district looked much like Rico's had in 1879. By the next year Lake City's first boom was evidenced by four assay offices, nine saloons, ten lawyers, two banks, three restaurants, fourteen general merchandise and clothing stores, and a variety of other businesses. When the boom passed in the late seventies, Lake City's specialized businesses decreased in numbers; for example, only two assayers and one bank remained. The appearance or disappearance of specialized businesses was one of several hallmarks of a camp's prosperity.

A closer examination of these business districts reveals that each—Lake City, Silverton, Ouray, and Rico—had or would acquire smelters, banks, and sawmills. The last was the most significant industry initially; a reporter for the *Dolores News*, December 13, 1879, wondered how Rico had advanced so rapidly, considering the continual shortage of lumber. Further refinements would appear, such as a shingle mill and, for some communities, even a brick kiln. Frame construction quickly superseded log, and for the wealthy brick came to replace frame. This, of course, improved the visual appearance and gave the desired semblance of permanence. Less progressive outlying camps either continued with log construction or paid the higher price for lumber. Rudimentary banking could be carried on by the general merchant, utilizing his safe or strong box; in the really progressive camp, however, a regular bank was essential for financial transactions and as a symbol of solidarity.

The cry of "fire" would be heard many times in the crowded, wind-and-weather-dried wooden camps, virtually explosive tinderboxes. In the 1870s only Lake City suffered serious damage, when an early morning fire on November 8, 1879 left the greater part of the business district in ruins. Mismanagement and confusion in fighting it aggravated the catastrophe, which served as the impetus for forming a fire department and equipping it with usable fire fighting apparatus. Lake City rebuilt with more permanent construction (brick buildings served as excellent fire walls). Far too often it took this kind of tragedy to motivate the town fathers and citizenry sufficiently to provide protection against future blazes. Fire showed no discrimination—the two-story business block went up in smoke along with the tackiest shack. Lake City was so indifferent that the large metal triangle used as a fire "bell" had been broken the previous spring at a chivaree, and no one had thought to repair it.

Professional people as well as merchants settled in the San Juans, generally in the four camps already mentioned, because of the larger clientele. It was no accident that each of these became the county seat; thus, lawyers congregated near the court. The San Juans were no different from any other mining region, and law suits came thick and fast over claim boundaries, ownership, and a multitude of other matters. Some lawyers drifted as much as the prospectors, one landing in Rico after sojourns at Silverton and Ophir. A doctor arrived in the same camp after hanging out and taking down his shingle in Silverton, Lake City, and Animas City, all since his arrival in 1875. Generally, however, both groups proved much more stable. The earliest arrivals tended to be more transitory, either because of professional incompetence or lack of work. Dentists were rare, usually of the visiting variety, who set up shop for a couple of weeks to fill and pull aching teeth.

Rico, Silverton, Lake City, and Ouray also cornered the few churches that were built during the 1870s and most of the schools, thereby increasing their general attraction for families and promoting that aura of stability so gratifying to the local boosters. Circuit-riding ministers toured the outlying areas, perhaps establishing a Sunday school class or corralling the nucleus of a congregation to be built upon later. It was not that the smaller communities did not want a church, school, resident doctor, or lawyer—they simply lacked the population base and were thus pushed by default into the zone of influence of their more successful rivals.[4]

Even at this early stage, a semblance of culture and dignity was extremely important to the local civic image. For this reason alone, some residents supported the establishment of schools and churches, especially the construction of buildings which improved the municipal "sky line." This kind of short-sighted support often failed to maintain or fund the institution once it was established, the building itself being more important than what it housed.

A discouraged Presbyterian minister wrote to the *Rocky Mountain Presbyterian*, September 1878, about his tribulations at Animas City. "The object of many seemed to be just to get a church building here as an addition in building up a town, and now when I insist on membership first, they have lost all interest . . . Many do not wish to give up dancing."

The *Silver World* warned Lake Citians in January 1877 that no money remained in the treasury to pay teachers and current running expenses. Two months later with teachers' salaries three months in arrears and the school owing local businessmen $500, the school board had no alternative but to levy a tax, an extremely unpopular measure. A dance to raise money netted $36, "a mere drop in the bucket," the editor noted. Mining camp denizens abhorred taxes of any sort, a fact city fathers had to keep in mind constantly during their deliberations. In the transitory world of mining it was much easier to levy taxes than to collect them.

The church suffered similar adversity. Much of its success depended upon the minister; if he were willing to take the congregation as they were and meet them halfway on various issues, then he stood a chance of acceptance. Too much religious doctrine, eastern snobbishness, or just plain stand-offishness could cause the flock to melt away. Fortunately, the San Juans were blessed with some outstanding ministers in the 1870s, though even the best never guaranteed that a church could be sustained. George Darley, one of the first rate and earliest, rode horseback and walked throughout the region from his base at Lake City. One had to "tire and tire again," he admonished, and be willing to preach in any building available (including saloons), face disappointments from one's "Christian brethren," and continuously battle financial crisis. Darley's "peculiar friends"—gamblers, prostitutes, and "drinking men"—quite often supported his endeavors better financially than did the so-called "good people," who had less ready cash than the first three groups. Many local churches stayed afloat because of their denominational home mission programs; their eastern brethren believed that the heathens out west needed saving as much as those in foreign lands.

Joseph Pickett went to Silverton in 1878 and found a situation somewhat typical of many other San Juan camps. Finding no competition, he conducted a religious service in the school with a dozen people present, including three women. A couple of weeks of meetings followed, producing a growing "interest in spiritual things." By the end of his stay he had nearly every child in town in Sunday school (local gamblers donated the money for a library), a full house at his services, and what he hoped was a foundation for later church organization. Pickett believed that an ecumenical church was the best for a camp. It was impractical to sustain several Protestant denominations in such small communities so distant from their neighbors. Regrettably, but understandably, most preachers and laymen disagreed and the 1870s spawned numerous Protestant congregations which sabotaged one another with proselytizing, denominational jealousies, and local scraps. The Catholics, unchallenged by these problems, were not much stronger, most camps being ministered to by a circuit-traveling priest, if they were served at all.[5]

The church and school, though small in number, had much to contribute to the cultural, intellectual, and spiritual life of the community. These bulwarks of Victorian America provided education, music, and lecture programs, acceptable social outlets, and an upgraded moral tone. The church could be especially effective in this latter sphere, if the minister adapted himself to the situation. Darley knew how to do it, as shown in this excerpt from one of his sermons:

> Sinner, whether you are a mining sinner or a prospecting sinner, do you wish to be "staked in" on this "lode," and have your name recorded in the Book that our Creator keeps, in which are written the names of all who are interested in it? If so, go to Christ; tell him you have thus far sought the gold and silver that perisheth with the using—the "veins" of silver and gold—but now you desire an interest in the imperishable riches . . .[6]

Women found the church an excellent outlet for their talents, and they generally became its

To get from here to there in the rugged San Juans was never easy. This crew is building the narrow road between Silverton and Ouray. Some tourists today would testify that the road is not much better now than then.

backbone through their attendance, work, and interest. The church and school occasionally joined forces to lead some reform fights, often involving the same individuals from both. In the 1870s the church seemed to fight the good fight more often, particularly for temperance (much pledging, much backsliding) and against liquor's handmaiden, the saloon (little headway).

Various fraternal lodges also came into being, providing a haven for brotherhood and enhancing individual importance in the transitory world of mining. But even in the fraternal world troubles lurked. The Masons at Lake City had a difficult time organizing in 1877. One "brother" went to Denver to "post" himself on the work needed to get a "dispensation." He was not heard from again, and eight months later the local Masons were still trying to find out the procedure for establishing a lodge.

Other cultural and social outlets arose, offering the illusive facade of civilization. A miners' or subscription library took the place of the public institution; neither money nor initiative was available to support the latter. A dramatic club, literary society, social club, or lecture series might flourish for a season, usually during the winter, when everyone had more leisure time. They faced a precarious existence, depending upon the originator's forcefulness, the continuing leadership, and interest generated among at least a handful of members. As might be expected, these activities were concentrated in the larger communities. Despite grand pretensions and generally good newspaper coverage, they failed to persuade more than a minority of the locals to become seriously involved.

Far more sought relaxation in the ubiquitous saloons or in the small but growing red-light districts. These men, the ones Darley described as "not rough characters," preferred masculine entertainment and recreation. The saloon well satisfied this craving in an atmosphere of male conversation and relaxation. Here also could be found Darley's "hard cases," a small handful who managed to cause enough trouble to taint the mining frontier with a reputation that most communities sought to live down or avoid. Gambling, drinking, gossip, congeniality, a place for conducting business—the saloon provided them all—and even art appreciation in the form of the reclining nude hanging over the bar. Obviously, no respectable woman would consider going near such places, unless to drag a husband home.

Two seasons were set aside just for relaxation and fun, Christmas-New Year and July Fourth. The early celebrations, though crude by later standards, involved wide participation and hearty enjoyment, simply because most everybody knew each other and there was little competition from commercial entertainment. Dances (a shortage of women proved to be no handicap), parties, school and church programs, and St. Nicholas's appearance kept the winter holiday full, while booming anvils and pistols ushered in a patriotic Fourth, featuring a program of speeches, noise, and plenty of toasts, perhaps topped off with a dance. The rest of the social season, generally delimited by the winter months, also included a number of dances or balls.

Throughout the seventies a scarcity of women prevailed, to the point that two young men, tired of enduring the "isolation of a miner's life," advertised for correspondence with a "limited number" of young ladies, the object being for the amusement and mutual improvement of both parties. When romantic dreams became reality with the arrival of members of the fairer sex, it could be quite an occasion. Dave Day printed this letter in his *Solid Muldoon*, recounting just such an arrival in the Sneffles District:

> Rinker put on Shultz's clothes and Shultz, in order to conceal the ventilation in his overalls, sat down on a rock and looked wise. Clark the "Terrible" took to the brush, and Mike Dermody sought refuge in the Yankee Boy Tunnel . . . These ladies were the first ever to visit the mines of Mt. Sneffels.

A bit overstated, perhaps, but the boys were eager for female companionship. The ladies who did come were usually married; the few singles seldom stayed that way. Some women combined homemaking with operating businesses, such as restaurants, dressmaking shops, and even a dairy.

In 1877 Colorado voted on the question of woman suffrage. The Lake City *Silver World*, at first unsympathetic to women, came out strongly in their behalf that year. In an editorial on January 13 favoring adoption, the editor wrote, "We believe that the time has come in Colorado for conferring the right of suffrage upon women." The *Ouray Times* was more conservative, urging the voter "to prepare himself as fully as may be." The time for woman suffrage had not yet come, however, and when the votes were tallied it went down to defeat statewide, badly so in Ouray, San Juan, and Hinsdale counties. The men voted it down in spite of the foray of the well-known Susan B. Anthony into the San Juans. A storm canceled her appearance in Ouray, but she continued undaunted to speak in Lake City to reportedly the largest audience up till then assembled there.[7] The locals listened, saw no apparent reason to extend the franchise, and promptly voted their conviction.

Though few in number, the women were beginning to make their presence felt. They pushed for reforms, they made homes in the bachelor's masculine world, and just by being there, gently smoothed the rough edges of the mining frontier. Woe to the husband who subjected his wife to mistreatment. She did not have to put up with it, as the high number of divorce cases testified. The court in Ouray listed seven for the September term in 1879.

The mythical average San Juaner faced a life much like his counterparts elsewhere, packed with hard work, disappointments, and those every-day incidents which could enliven otherwise ordinary routine. The cost of living stayed high, thanks to a combination of long freighting distances, high freight rates, and the disinclination of the ordinary miner to be very self-sufficient. Alleviating some of the problems was the gradual settlement of farmers near those camps that had land available at lower elevations. This meant Lake City and the Animas Valley, primarily, with Ouray and Rico also hopeful of some farming success. Valiant attempts were also made at Howardsville and even Imogene Basin, high in the Sneffels district. This latter experiment simply involved seasonal pasturing of cows and selling milk and butter, a not unprofitable venture.

Merchants showed no aversion to taking advantage of their situation; George Darley commented that even needles cost ten times what they did back east because the freight was "so high." Everything was high, not just the altitude, and in the late winter and early spring seasonal price increases reflected dwindling reserves on merchants' shelves. In 1875 a shortage of money compounded business problems, particularly in the Silverton area.

Regardless of what some of his customers might think, the life of the merchant was not a bed of roses. After a wearisome struggle crossing snowy Stoney Pass with eleven steers, Englishman John Scrivner opened a butcher shop at Howardsville in May 1874. "We had no paper or string, and the people carried away their beef on skewers." Eventually satisfied that a living could be made, he and a partner built corrals, a slaughter house, and a cabin. "We had a pretty good business and made a little money." For Scrivner and others, overhead costs were high and business potentially risky, and, in addition, they faced the wrath of customers outraged by high prices.

Another annoyance was poor mail service, especially since every camp with any pretense (and there were none without) wanted nothing less than daily deliveries. Snow and storms did manage to stop the U.S. Mail; at times, only the carrier's determination got it through. In the winter he donned skis (called snowshoes then) to try to deliver the mail to snowbound camps; in the spring he fought swollen streams and mud. Heroic efforts, though locally applauded, did not quell the rising tide of protests. To a lesser degree the lack of standardized time annoyed both visitors and natives. James Hague observed that, while his watch said 10:30, local Silverton time was 11. Bothersome at the least, it could be downright inconvenient if one happened to miss the stage which was running on a time different from one's watch.

The San Juaner perchance could afford to be late, but neglecting his health could prove disastrous. Death lurked in every mine, along the trails, and in the towns. Accidents took a grim toll of maimed and dead. Lucky were those who survived a snowslide; one who did recounted his experience. Starting to walk from his cabin to the mine about 100 yards away, he suddenly found himself in trouble.

> A slide started at my feet and I went down the hill with it, rode it or it rode me, to the bottom. I was badly bruised. I thought my back was broken. My clothes were packed full of ice. When it stopped I was sitting up on the slide and could not move. I lost my hat and ruined a new pair of boots . . . I went over a bluff, must have been 50 or 60 feet sheer, perhaps more.

Pneumonia, the dreaded killer of these elevations, might begin innocently enough as a cold or sore throat. Cold, drafty cabins offered little protection for those trying to prevent illness, nor were winter diets conducive to good health. Too often no doctor was near at hand and home remedies proved inadequate to stem or cure a minor ailment before something more serious set in.

Free medical advice, among other things, was offered by the newspapers. Chronicles of the miners' lives and all that went on around them, they represented a vital force, the San Juaners believed, in their district's future. And in many ways, they were—town booster, defender, advertiser, and social, political, and local gadfly. Along with the church and school, they were symbols of respectability. Lake City published the first newspaper, followed by Silverton, Ouray, Rico, and Animas City, showing again the predominance of the big four. The editors recorded many firsts, like the first piano (definite cultural aspiration), the inaugural band concert, the largest homegrown vegetable, the first murder (embarrassing to a camp's image), and the first-born baby (who usually received a variety of gifts). The editors advocated civic improvements, such as street cleaning, watering down dusty streets, and saving local trees (early ecologists), and were never remiss in flaunting local attractions—Ouray's mineral springs, for example. They offered advice on the advantages of shingle over mud roofs, what businesses were needed, the benefits of patronizing local merchants, especially the newspaper, and how to promote the camp.

In this last category they stood in the vanguard. Lake City was number one to the *Silver World*, Ouray to the *Times* and *Solid Muldoon*—and God help the person or newspaper who challenged local chauvinism. Like young kittens spitting and hissing, the newspapers fought back and forth, sometimes eliciting a hurt cry with an especially painful slash. Rico maligned Lake City because of its high freight rates; Silverton liked it no better because of the poor mail service. Lake City lashed out at Ouray as an "unmitigated fraud and humbug" with no future, to which the *Solid Muldoon* responded by describing Lake City as on the "low grade" end of the San Juan. Rico, in Ouray's eyes, was a "sinful camp"—and so it went. Lake City received the brunt, its head start guaranteed to arouse local jealousies. These intradistrict fights were within the rules, but let an outsider join the fray and local newspapers would rush to each other's defense in an amazing show of unity. San Juaners could attack each other but no outsider need cast aspersions.

A transitory lot were those early editors, willing to put up with the same privations and primitiveness as their customers. Their papers crammed into four pages the local, state, and national news, and plugs for local businesses, numerous advertisements, and probably an editorial. Uniformly, the first issue announced their goals, primarily to devote themselves to local interests and to inform the outside world of riches that were begging to be developed. Occasionally the editor promised to preach no sermons or dabble in politics—a little too much to expect.[8] Out of the nine newspapers begun, two had folded by 1879.

John Curry of the *La Plata Miner* was typical of these pioneering editors. After tramping over Stoney Pass in 1875, this Iowan visited Howardsville but decided to cast his lot with Silverton. This proved to be one of the fateful decisions in the rivalry between those two camps.

Ouray and the other mining communities became the supply points that allowed year-round mining. Burros could eke out a precarious existence on the scant grass in the mountains, but mules required feed. Both were scavengers of the debris scattered carelessly about these communities.

Curry's hand-operated equipment was dragged over Stoney, and the first issue of the *Miner* appeared on July 10. Curry did not winter in Silverton, returning home to his family instead; an assistant ran the paper. While back in the midwest, he vigorously promoted the San Juans, always coming back to Silverton in the spring. Finally, in 1877, he felt that the town had developed enough to bring his wife, who returned to her home in Sandwich, Illinois to have a baby in 1879. Curry's refusal to winter in Silverton and his desire to ensconce his wife in Illinois for the birth testify to the still primitive nature of Silverton.

Most of the political action occurred on the local scene. The only real test of regional political sentiment came in 1876 and 1878 with the first poststatehood elections. The Republicans pulled out a narrow victory each time. Although in the most isolated part of Colorado and unable to wield much political clout, both parties organized themselves for a partisan effort, making arrangements to get out the miners' vote, endeavoring to raise money, and sponsoring debates between the candidates for the single U.S. House of Representatives seat.

Representing the San Juans in the 1877 legislature were one senator and four representatives, three of whom gave mining as an occupation; one was a merchant and one failed to list anything. In the 1878 election the Republicans captured the single senate seat and two of the four house seats.

On the local level the men, and sometimes the issues, retained importance equal to party labels. When town governments were just being organized, vested interests, old grievances, and poor performances had not had time to play the role of spoiler. Local elections were looked upon more as "firsts" than as heated political campaigns. Partisan politics, in their infancy at best, showed none of the refinements of Denver, let alone an older eastern town.

In 1874 the San Juans were divided into La Plata, Hinsdale, and Rio Grande counties. Two years later San Juan County was carved out of La Plata, and the next year Ouray was created from the northern portion of San Juan. This reorganization should have satisfied local politicians, but it failed to placate everyone. Angered over being ignored, and in a state dominated by eastern slope interests unacquainted with the San Juans, some malcontents agitated in 1877 to create a new territory, named appropriately enough, San Juan. Emphasizing the lack of common interests, the few unhappy souls were nonetheless unable to lead anyone out, and the idea died aborning. Miners were too busy mining and speculating to be deterred by statehood hurrahs.[9]

Many more people showed interest in creating camp governments, and here were seen the most concentrated political efforts. With people thronging into congested sites, urban problems quickly multiplied and forced consideration of the necessity for municipal government. Aligned against each other were opponents of the inevitable expense involved and those who strongly doubted the need versus those who touted the benefits of governmental agencies and emphasized the need to handle current issues beyond simple stop-gap remedies. Each San Juan community wrestled this one out for itself. Most decided not to go ahead, but the larger ones were ultimately forced to do so.

Local residents believed the less government the better and expressed no wish to be burdened with additional taxes. Out of this sentiment grew the idea that an elected board of trustees would be the best solution. The board, in turn, would hire as many officials as it deemed necessary to provide adequate and economical services. Silverton illustrated what was wanted and what was secured. The board was required to meet at specified times each month and could call special meetings when necessary (nine from November 1876-April 1879), at which time they were supposed to discuss and vote on a published agenda. A quorum was necessary to transact business, a requirement that nullified eight sessions during the above-mentioned period. During its meetings, the board passed town ordinances, listened to complaints from the citizenry, selected and replaced town officials, appointed standing committees (by 1879 there were five, including finances, police, streets and alleys), purchased promotional pamphlets, and handled routine matters. A flood of reports also crowded the agenda. Initially the agenda were very full, interest was high, and much time was spent on writing ordinances. Enthusiasm quickly waned, especially in the winter when the meetings were short or nonexistent; none is known to have been called from November 1, 1878 to March 8, 1879.

Petitions from Silvertonians aired a multitude of grievances or promoted special interests. One group entreated the board either to suppress dance houses or restrict them within certain boundaries (discussed and defeated). Another group wanted fire equipment purchased (approved). In a related matter, petitioners asked the board to construct water ditches for fire protection (approved). The local rifle club petitioned to establish a rifle range; it was approved, provided the club held itself responsible for any damage. And so it went, as the residents and their board hammered out governmental services and ordinances that responded to necessities of the moment.

Ordinances were passed with alacrity, then repealed or amended as the need arose. At first, ordinances from other towns were simply copied or modified to meet unique circumstances. Silverton's first five created a town seal, explained the town officers' duties, established business licenses (the main revenue-producing source), defined violations of breach of peace, and, last but not least, specified how to initiate and pass ordinances. Each ordinance, as well as legal notices, had to be published in the local newspapers, a requirement that often meant the difference between profit and loss for the struggling owners of the paper. Sharp competition developed over who would get the town's printing bid.

The governmental offices most relevant to the average resident were the town clerk, who carried on the day-to-day business, and the town constable (title varied from camp to camp), who enforced the ordinances. Silverton's town clerk was also treasurer and was required to post a bond of $1,500. These two, joined by the city attorney, street commissioner (quite often the constable), justice of the peace, fire warden (1877), and night watchman (1879), constituted Silverton's officials. Practicality and economy were noteworthy throughout. Jobs were combined or abolished and salaries even reduced; for instance, the constable's was $125 a month during eight months (apparently a busy time) and $75 from December to April. At one time it had been $75 and $35, but qualified men were hard to find at that hardly livable wage.

A constable had the most abrasive job, which included everything from corralling camp drunks and prohibiting minors under eighteen from frequenting or loitering near saloons to capturing stray dogs and making arrests. Silverton had a difficult time securing and keeping a man who fulfilled everyone's expectations.

The population did not wake up each morning expecting to be assailed with news of some criminal misdeed, nor did the San Juaner feel naked without his gun—he was suspect if he wore one. If the Lake City paper can be trusted, the first killing in that camp did not happen until July 1876. The major problem was simply trying to find a jail strong enough to keep the prisoners (generally drunks) within. Rico, on the other hand, confronted its initial murder within a year of its founding. Burro and horse stealing and lot jumping were more common serious offenses. Many early problems were handled privately in an extralegal manner, such as the case involving a man who skipped out of Animas Forks owing the lady boarding house operator three months' rent. A group of the "fellows" traced him to Silverton, surrounded the culprit, and announced, "You got to come back and pay your board bill or give us the money, or we'll hang you." The money was promptly found and the fugitive fled, never to return.

Lawlessness increased at a rate commensurate with the number of people arriving. The formation of the San Juan Detective Association in September 1878, for mutual assistance in criminal detection, reflected a new need. The boys got off to a congenial start with a rousing champagne supper at Silverton, which heightened spirits if not acuity. Criminal activities could not be tolerated—they sabotaged all other efforts to convince outside investors that the San Juans had passed through the rowdy frontier days. Local newspapers were aware of this and may have been guilty of playing down what criminal activity actually occurred.[10]

Limping along for three years without an adequate financial base, the Silverton board finally, in July 1879, instructed the town attorney to prepare a tax ordinance on real and personal property. At this time the main expenses included the clerk's salary at $25 a month, constable at $125, the attorney, who was paid $300 per year, and now also a treasurer, being paid $200 per annum. Incidental expenses for publishing legal notices, renting a room for council meetings, boarding prisoners, and the like, also mounted steadily, despite all attempts to keep costs down.[11]

Silverton, like other San Juan camps, found that once the government got started, it grew inexorably, expenses and regulations proliferating with it. Urbanization here was no different from elsewhere. Though the pioneering situation might have magnified or repressed certain issues, they were still present and called for solutions. Finances, rowdiness, and street maintenance were handled by whatever means were available. Numerous ordinances were passed that remained only deadwood on the statute books, but they gave an impression of municipal awareness if not responsibility. In the awkward years of birth, mistakes were made, to be sure, but a government was forged and functioned to the satisfaction of the majority—or they turned the "rascals" out at the next election.

As this decade came to a close, at least one festering problem was concluded to the San Juaners' satisfaction. By the 1873 Brunot Treaty, the Utes had been removed northward out of the San Juans, a solution that was only temporary. Ouray (named after the Ute

Lake City was a mining town, not a camp. Its size, and the brick and stone architecture, are shown in the photograph. A town also possessed a larger business district, more wealth, and a different community spirit from a smaller camp.

leader) was situated only a few miles from the reservation and its best natural outlet ran right through it. This would have proved intolerable under the best of circumstances, even if there had been no fear of the Indians. James Hague, among others, expressed that fear in 1876 when he wrote that he had been unable to visit the Mancos River Valley (near the Southern Utes) because the condition of the Indians, while not positively dangerous, was such that it was deemed prudent not to risk an extended stay.

The Ouray papers did much to incite anti-Indian sentiment, undoubtedly mirroring their readers' attitudes. As early as 1877, the editor of the *Times* questioned why the Indian reservation covered some of the best land in the region, thus preventing its occupation by "intelligent and industrious citizens." By February 1879, the stand became much more adamant: "The Utes Must Go." Ouray was joined by Lake City's *Silver World*, also uneasy about the nearness of the Utes. Pressure started to mount locally, as evidenced by petitions to Congress for removal. A fascinating dichotomy evolved: the Utes had to go but Ouray himself was acceptable, and his namesake camp turned out on March 9, 1879, to give him a royal reception when he came for a visit. The *Solid Muldoon* also praised the Utes for vegetables they had grown and were selling that fall; this was fine—it reflected adoption of the white man's ways.

The inevitable happened. The two cultures clashed, resulting in the killing of Nathan Meeker, Ute Indian Agent at the White River Agency in northern Colorado. Although fairly distant, this outbreak engendered a pervading fear of raids, and the San Juaners had had enough. They responded predictably by calling for troops. This expedient was not fast enough for some, so they organized local militia units, putting up a bold front at least. No Indians swooped down, nor did war whoops echo through the canyons, and the fear subsided. The rest of the state joined the San Juaners—pressure was put to bear on the government— and within a few years the Utes were gone, hustled over to Utah. The last chapter of the Utes and San Juan mining was closed, the land north of Ouray now opened to white exploitation. Only the numerically weaker southern bands remained, and they were well outside the immediate mining districts.[12]

While rumors of Indians did not materialize for most San Juaners, neither did the wealth they anticipated. Pioneering in the San Juans was hard work and a day-to-day existence. This was no more clearly shown than in the 1876 diary of T. Parker Plantz, who was then living between Silverton and Howardsville. Miner, doctor, sometime Democratic politician, fifty-two-year-old Plantz jotted down each day's activities. He had come to Baker's Park the year before, perhaps to escape an unfortunate second marriage. His mining activities involved assessment work on his numerous prospects, which produced little more than his own hard work. Here are excerpts from his diary (spelling and punctuation have been left intact).

[July 4]
Beautiful bright day and night. I went down to Silverton this morning to attend the celebration of the fourth, being the centennial fourth of the United States. Having no canon we substituted anvils, they answering a very good purpose. After the reading of the declaration of Independence we had a short but spicy address delivered by Judge Burroughs . . . where I had a good view of the races. After which I took supper at the Earl hotel. I strolled around different places until about ten o'clock. I went to the ball, enjoyed my self very much. Feel well.

[Aug. 22]
Clear and pleasant all day. John and I went up to the Round mountain basin where our tent is arriving there about noon. Had our dinner, after which we went up to the Christian lode and finished the assessment on it, finding no mineral yet, but there is good indications of mineral when we get deeper down, which we will do next season. I called over to O'Neil's tent in the evening having a long talk about machinery and minerals in general. John cooked some beans this evening. It is clear and cold to night. I am feeling well, but tired.

[Sept. 14]
It was clear and warm this morning, but it turned quite cold about noon and clouded up and threatened to rain, but finely cleared off and became warm. John and I have been working on the Prize lode all day. We lowered the dump about three feet and prepared a place for piling up mineral. We have done a good days work taking out a quantity mineral. We put in two shots which did good execution. L.W. Chapman has been here this evening, talking over the mining laws, also the political issues of the day. It is clear and pleasant tonight. I am well but tired.

[Oct. 17]
Morning clear gave the appearance of being settled weather, but about the middle of the day clouded up and commenced to snow quite hard, upon the mountain. John and I being at work upon the Croghan, the snow bothered so that we had to quit and come down to the cabin, we done a good days work taking out a large quantity of rock and mineral. Mr. O'Neil called at the cabin this morning also this evening. This morning two of the fancy girls went out. Several parties went out today, the snow did not lay on the ground in the park. John & Isaac is settling up. I am feeling well to night.[13]

These were the events that illuminated, and perhaps altered, Plantz's life in 1876. Around him the San Juans matured.

The decade of the 1870s brought substantial urbanization to the San Juans. Pioneering in a mountainous isolated land with an optimism natural to the American frontier, town builders came on the heels of the miners. Within these few, fast-paced years settlement and a claim to permanence had arrived. Times were never easy, dreams had turned to nightmares for some, and the best of plans had a way of proving completely impractical. Still, a foundation had been laid and the urban San Juaner could look forward with justifiable optimism to the 1880s. Certainly those coming in the next decade would find their way much easier because of the individuals who had come in the years now past.

[1] Jones to Georgina Frances Jones, August 3, 1879, Jones Collection. D.W. Brunton, "Technical Reminiscences," *Mining and Scientific Press*, November 27, 1915, pages 6-7. Hafen, *Jackson Diaries*, page 304, Hague, Parrott City Notebook, September 16-18, 1876.

[2] For information on Silverton see Camp Journal, August 2, 1875; *Rocky Mountain News*, October 8 and November 18, 1874 and September 22, 1876; Sara Titus (ed.), *This is a Fine Day* (Long Beach: Seaside Printing Company, 1959), pages 47-51. For Howardsville and Eureka see *Rocky Mountain News*, February 19, November 1, 18, 1874, June 30, 1875 and October 17, 1877; *Silver World*, July 31 and October 16, 1875 and August 26, 1876.

[3] Hafen, *Jackson Diaries*, pages 305-324. *Rocky Mountain News*, October 23, 1874, November 25, 1875, May 17, 1876, and March 31, 1877. Captain John Moss Papers, Colorado Historical Society. Henry Williams, *Williams Tourists' Guide* (New York: Henry T. Williams, 1877), page 12. See also Parrott City Notebook, Hague Collection.

[4] The previous section on San Juan large and small camps was taken from the *Rocky Mountain News*, 1874-1879, *Silver World*, 1875-1879, *Dolores News*, 1879, *Ouray Times*, 1877-1878, and *Solid Muldoon*, 1879.

[5] George Darley, *Pioneering*, pages 17-22, 53-57, and 67-68. William Salter, *Memoirs of Joseph W. Pickett* (Burlington: George Ellis, 1880), pages 94-99 and 107-111. See also newspapers listed in footnote 4 for day-to-day accounts.

[6] Darley, *Pioneering*, page 115. See also the Darley Family Collection, Western Historical Collections, University of Colorado.

[7] *Ouray Times*, July 14, September 2, 21, October 29, 1877, July 27 and August 3, 1878. *Solid Muldoon*, September 5, 26, 1879. *Dolores News*, September 23 and October 25, 1879. Darley, *Pioneering*, page 114. For the Lake City Masonic troubles, see Henry Olney to Ed Parmlee, April 1 and August 22, 1877, Orahood Papers, Western Historical Collections, University of Colorado.

[8] Not all the San Juan newspapers had long runs or even many copies still in existence. For the 1870s see footnote 4.

[9] *Ouray Times*, October 6, 1877, discussed the Territory of San Juan. Frank Hall's *History of the State of Colorado*, volume 4, traced county development. Thomas Corbett, *The Legislative Manual . . .* (Denver: Denver Times, 1877), pages 228-230 and 331-343. The William Hamill papers contain a few San Juan items for 1878. The newspapers remain the best general source for each election.

[10] *Silver World*, August 7, September 25, 1875, April 29, May 6, June 3, July 8, August 26, and September 23, 1876 and May 18, 1878. *Dolores News*, September 4, 11, November 15 and 22, 1879. *Ouray Times*, June 30, 1877 and September 21, 1878.

[11] Silverton, No. 1 Minute Book Board of Trustees, November 18, 1876-December 31, 1879. Ordinance Book-A, 1879. Silverton is the only San Juan camp from this decade with any large collection of records.

[12] Hague to T. Parrott, October 12, 1876. Hague Collection. *Ouray Times*, November 3 and December 22, 1877, February 8 and March 15, 1879. *Silver World*, April 27, 1878. *Solid Muldoon*, September 26 and October 5, 1879. *Dolores News*, October 25 and November 8, 1879.

[13] T. Parker Plantz Diary, 1876. Copy in possession of author.

A
Time
to Wait

Few of those who stood on the threshold of the 1880s could have correctly forecast what this decade held in store for San Juan mining. These were to be the transitory years between the opening excitement and the final fulfillment—a time of individual dreams turning sour, of might-have-beens becoming used-to-be's. Like wispy smoke drifting from a cabin's chimney, many of the "mines" evaporated in a twinkling. Those that prevailed kept the miners searching. The mining districts proved just as contrary, marching at their own pace; each went its own way, often oblivious to all effort and planning.

On the whole, the San Juans made advances; total production mounted steadily. Horatio Burchard, director of the mint, was amazed, calling it in 1883 "beyond reasonable expectation." Individual production patterns, however, were kaleidoscopic. Rico, the youngest district at decade's start, went from mild boom to bust and back to boom in 1889. Exuberant supporters hailed it as the second Leadville. Silverton, the oldest district, solidly improved its output, topping $1,000,000 per year in the mid-1880s, before leveling off in the high hundred thousands. Summitville soared and then collapsed, barely beating out La Plata County for last place in yearly production in 1889. Even with some early promise from La Plata Canyon and nearby mines, La Plata County's total gold and silver production never equaled that of even a mediocre mine, and it retained its solid hold on last place. More encouraging in that area was the emergence of coal mining, which held real potential. Thanks to the new discoveries on Red Mountain, Ouray jumped its output to the million dollar plateau, where it remained throughout the eighties. Hinsdale had its troubles, first holding steady, then slumping, only to recover for a couple of years before sagging again—a jerky pattern that cost the county its number one ranking. Telluride, as it was now officially called, shifted interest from placer operations to quartz mines, which came to dominate, and whose output in 1889 lifted the district over the $1,000,000 mark in gold and silver.

Out of this jumble of excitement, prosperity, recession, and despair emerged one unmistakable trend—the increasing dominance of the principal San Juan districts: Telluride, Ouray, and Silverton. Others might seriously threaten to move ahead, only to peak and recede. Investors should have put their money here, but foresight is a much rarer commodity than hindsight. Not so obvious in the foregoing was the San Juans' emergence by the late eighties, after a wait of nearly two decades, as one of Colorado's major mining areas. Even so, it was still overshadowed in the public eye by more glamorous districts; its day had not quite arrived.

The earliest significant development in mining came not from increased production but from the means to that end—the railroad. The road which logically tapped this mountainous mecca was Colorado's Denver and Rio Grande (D&RG), a pioneer in reaching a number of mining districts. As far back as 1875, its guiding genius, William Palmer, had cast longing glances toward what he described as "more rich silver lodes" than had been discovered elsewhere in any western state or territory. San Juaners proved willing to be pursued, only to see the courtship languish temporarily because of Leadville's more glittering attractions, railroad rivalries, and financial difficulties. Palmer never lost his enthusiasm, though, and by 1881 the road was well on its way. It reached Chama, New Mexico in February and pushed northwestward on the San Juan Extension in spite of winter weather. Animas City was the ultimate target of the onrushing railroad and gleefully anticipated its role as a railroad center. Assuming they held all the cards, the Animas Citians balked at meeting Palmer's terms. When the dust settled, the D&RG had established its terminal at Durango, only a couple of miles south of its rival down the Animas River. Animas City was soon surpassed, its railroad-backed neighbor greedily swallowing up most of its business and residents and grabbing the cherished designation as county seat. By the time the tracks reached it in July, several thousand people had crowded into Durango's jerry-built site. The local celebration had hardly subsided when the D&RG pressed on up the Animas Valley toward Silverton, arriving there in July 1882. Animas City, now merely a stop between the two, could only ruefully contemplate what might have been.

After a respite the Denver and Rio Grande, in 1887, added an extension from its main line at Montrose, southward thirty-six miles into Ouray. Then at great expense, to hold off a rival company, it built into Lake City in 1889. One setback marred the railroad's record: a short line from Del Norte to Wagon Wheel Gap, to tap this tourist spot, was discontinued in 1889, because considerable snow damage each spring cut too deeply into profits. The most active railroad in the eighties, the D&RG turned the San Juans into its exclusive territory, locking them into its economic orbit in return for providing the fastest, cheapest, and best (weather permitting) year-round service available. Besides these obvious contributions the railroad helped promote the region in numerous pamphlets, and the town to which it served as midwife, Durango, grew to play an important role in the regional economy.

The San Juans did not remain the Denver and Rio Grande's private fiefdom; the challenge came from none other than Otto Mears. Busy building toll roads into Telluride, Sneffels, Poughkeepsie Gulch, Red Mountain (via both Ouray and Silverton), and from Animas Forks to Silverton to replace the unsatisfactory earlier road, Mears nonetheless grasped the advantages of railroads.

Converting his Silverton-to-Red Mountain route into a railroad right-of-way, Mears opened traffic to Red Mountain in September 1888 and into Ironton and the park beyond the next year over his Silverton Railroad. Stage connections carried passengers into Ouray via his spectacular cliff-edge toll road. The railroad was an engineering feat, winding almost continuously once it started to crawl up the mountains. It ran two daily trains each way, dropping off freight at the spot nearest to the mine that ordered it. With connections at Ouray and Silverton to the D&RG (Mears could never hope to be free of their influence), new tourist possibilities for the "Rainbow Route" opened, something on which both railroads quickly capitalized.

These roads were all narrow-gauge (3 feet wide as opposed to the standard 4-feet, 8½ inches), a dimension Palmer and his associates had settled upon back in the early 1870s. Their decision, and that of Mears, was influenced by several factors; construction costs would be cheaper and the gauge more flexible, allowing easier maneuvering in the mountains, where expensive excavation, grading, and tunneling challenged engineering and escalated costs. Actually, Mears had no choice but to follow the D&RG lead if he planned to tie into their system, which he was forced to do in Silverton. Thus were the San Juans joined to the rest of Colorado and beyond by the narrow-gauge and its little engines.[1]

Mine owners on or near the tracks accrued immediate benefits, highlighted by a reduction in freight rates. William Weston pointed out that the cost of sending ore to Pueblo smelters had declined by early 1883 to $35 a ton, compared to $65 to $80 when he first arrived, now allowing the long desired shipment of lower grade ore. He also felt the railroad lowered the cost of living; however, the development of regional agriculture was at least partly responsible. Some disagreed with Weston, believing that the cost of living could be even lower if the

Denver and Rio Grande would reduce its "exorbitant" charges. The *La Plata Miner*, August 13, 1881, tore into the yet-to-arrive D&RG for supposedly trying to depress local mine prices (so Palmer and others could secure bargains) by manipulation of the smelter market. Despite such grousing, which continued on and off for years, there could be no question that the cost of coal, essential for mining operations, went down, as did overall expenses, provided one lived near the tracks. This last fact needs emphasizing, because those mines at a distance were still at the mercy of the freighters and their charges for hauling to and from the railroads.

Nearly year-round transportation came with the Denver and Rio Grande; at least there was better assurance of it than burros and wagons had given. But not even iron snowplows and human shovelers could keep the tracks open some winters, when slides ripped down the mountainsides and made it a losing fight.

Isolation ceased to be a drawback by decade's end, when the iron rails pierced the heart and embraced three sides of the San Juans. Instead of a week or better, connections with Denver and eastern cities were measured in days. With the telegraph and that new invention, the telephone, the San Juaner had the fastest service possible. Investors now could ride in comparative comfort almost to the doorstep of their prospects, no longer having to face the laborious travel that confronted Senator Jones only a few years before.

The railroad's impact was shown most dramatically on mining in the Silverton District, which had produced $97,000 the year the D&RG arrived, then multiplied that by four times to $400,000 the next season. The million-dollar level was surpassed in 1885 and, thanks to fortunate railroad connections, Silverton enjoyed a prosperous decade. Consider in contrast, however, railroadless Rico. Its plight caused a writer to complain to the *Financial and Mining Record* (New York), June 22, 1889, that its comparatively isolated position kept a great many people out and mining was handicapped through the absence of a railroad. Only twice in the eighties did Rico's mines produce over $200,000, lending the support of statistical evidence to swelling local cries for rail connections. Between these two extremes were the other mining districts, although it should never be assumed (as some individuals of the 1880s did) that the railroad was the magic elixir to cure all mining ills. La Plata County's mines illustrate this; though they were within five to twenty miles of the D&RG tracks, compared to Rico's mountainous fifty miles from Durango, they were simply not rich enough to support much production. When the ore was there, the railroad could help, particularly in marketing low-grade tonnage, which in bulk quantities could be the saving grace of a mining operation.

Across the San Juans the railroad's coming electrified mining interests and drowned out the few pessimists' protests that this could be a dubious blessing, one which was liable to clamp a heavy-handed grip on developments. Summitville found it no panacea and became the first major district to collapse permanently. The reason was not too difficult to ferret out: the miners burrowed past a zone of secondary enrichment to encounter irregular veins of low-grade gold ore which was exceedingly difficult to separate because of its chemical composition. With costs rising steeply, ore values plummeting, and no rich areas being opened, the large companies closed down, miners drifted away, and Summitville, a small area at its mining best, ended its major role in San Juan mining history.

Without a railroad Lake City, like Rico, suffered isolation pains, its low-grade ore unable to produce a profit against the high mining costs. Whatever early prominence it had enjoyed vanished in the eighties, the railroads harshly emphasizing this point by circumventing less productive and profitable Hinsdale County in favor of other San Juan communities. Where the railroads went so did investors and publicity, making Lake City a perennial bridesmaid. Somewhat brightening this drab picture were the Ute and Ule mines (worked as one operation), which continued to be the best producers, even going so far as to maintain regular shipments in the winter of 1887-88 to try to induce railroad construction. The year the

The San Juan operations pioneered in the use of trams, which could overcome the cost of freighting, bad weather, and mine isolation. Men rode on them occasionally, too. Engineer T.A. Rickard wrote, "The aerial voyage was made speedily and safely, if not very comfortably." He thought it wise to ride them in the winter and spring when the avalanche danger was high.

D&RG finally built a branch line (1889), total production sank to a low of $29,000; then, as in Silverton, it shot up, benefiting the next decade.

Symptomatic of local distress was the camp of Carson, straddling the Continental Divide southwest of Lake City, hardly a prime location (11,500-12,000 ft altitude) under the best of circumstances. What might have been years of promising development were blighted by exceptionally heavy snows and hard winters; narrow, irregular veins, with only a few small, rich pockets; and a steep, rocky trail to Lake City. This town contributed no help, but only its own troubles, to its satellite. Carson's silver mines were never developed; its youth passed almost unnoticed to all but a few.[2]

Enough of such depressing examples—these two districts were atypical of the San Juans. In Red Mountain's growth may be seen how high-grade ore, coupled with advantageous outside investment and better-than-average transportation connections, could be parlayed into the prosperity denied Carson. Red Mountain was the magic name in the 1880s.

Located between Silverton and Ouray at the southern end of a long, narrow valley, Red Mountain had first been prospected in the late 1870s. General attention, however, was not drawn to the area until the 1882 discovery and the opening of the Yankee Girl and Guston mines. Ore in both proved rich at the "grass roots," so rich in fact that it was not sorted, simply sacked and shipped out. A microscopic rush developed that spilled into the next year, with claims being filed on any likely looking spot. Ambitious, tenacious little settlements clung to mountain slopes, each claiming to be the heart of the district and vying for publicity by sending out almost daily reports of rich discoveries. With the Yankee Girl ore running at 2,000 ounces of silver to the ton, investors needed little convincing that any nearby claim would put them in "the lap of the gods." Both the Yankee Girl and Guston were sold before development hardly got started. They would eventually be resold to English investors, who controlled the major mines by the late eighties.

From 1884 Red Mountain produced steadily, especially once the mountain-locked isolation was overcome (primarily by Otto Mears, first with his toll roads, then his railroad). The Silverton Railroad appeared at exactly the right time; the high-grade ore was giving out, and it provided the means for shipping lower grade profitably. Red Mountain was fortunate—its own known riches and production record had spurred Mears on and he in turn boosted mining.

Unlike those of other San Juan districts, the ore bodies at Red Mountain were found in nearly vertical chimneys a few feet wide and often several hundred feet deep. The lucky individuals who tapped the right spot could find a bonanza in their laps; a next-door neighbor might find nothing, though his site seemed equally favorable. Both prospectors and investors were fooled. Fortunately, the ore remained rich after the first discoveries, because mining at Red Mountain encountered other difficulties. Water soon caused grief; not just ordinary water, but water so acid that it corroded every bit of iron it touched, eating through particularly exposed parts in only days. Charles Leonard, a worker there, tested steadily dripping water on a three-inch standard pipe and found that it took only nine hours to eat a hole through it. Expensive replacements and experiments to find resistant material cut into profits. Investors feared what might happen if the situation got out of control, as it did in 1888, when corrosion of a pipe caused the Yankee Girl to flood and stopped operations.

Another disturbing element was metallurgical: these ores were combined with copper and could not be reduced profitably by nearby smelters. Shipment to Denver or Pueblo was the result, until a copper-smelting process could be introduced locally. Through it all Red Mountain mining persisted. Its major mines were well developed, the Yankee Girl's main shaft down 965 feet by August 1889, and the district production continued strong throughout the decade. The geological formation dictated, however, that for every Yankee Girl, Guston, or Congress there would be many disappointments. But the lure in the eighties was such that prospectors and investors persisted, pumping fresh enthusiasm and money into claims hardly worth the time taken to file on them.

The Telluride area grew decidedly more quietly, though located just west across the mountains from Red Mountain. As in the seventies, exaggerated expectations outran reality. Attention was focused on placer and quartz mining; unfortunately, these anticipated twin pillars proved to be only weak reeds. Placer operations, especially, offered rewards far short of the effort expended by individuals who continued to see gold glimmering in the sands of San Miguel.

Spring after spring the eighties brought forth encouraging reports of placer projects that promised to pay unlimited dividends in the coming season. By fall, shorter reports told of dashed expectations and liberally distributed the blame. For example, the 1888 prospectus of the San Miguel Gold Placers Company assured stockholders that, beyond question, annual dividends of "at least" twenty percent would be paid. The company owned about 900 acres of gold-bearing gravel bars, which it considered superior for hydraulic operations, and promised to recover not less than $6,000,000 worth of gold. Swept away by its own rhetoric, the company pictured labor and supplies as abundant and cheap and the climate as mild, allowing more than a nine-month working season. This golden treasure box would be opened for the mere outlay of $100,000. Work actually started the next year, but by fall reports leaked out that the clean-ups were not satisfactory, three-fourths of the sand handled was barren, and the rest "darn poor." Expenses only climbed higher and were blamed on a small army of idle men languishing on the payroll, generally of the "kid-glove variety." To the rescue rode the old Civil War general, Benjamin Butler, who had a habit of dabbling in western mining. Elected president of this company, he announced that, without doubt, the company controlled a series of the country's largest placer mines, on which operation would begin at once. Stockholders, placated only by more promises, saw more outgo than income.

Left to placer operations, Telluride would have languished dismally; fortunately, the mines galloped to the rescue. Operations in the mountain basins surrounding the river valley were of a "blast and pray" nature; the miners were chronically short of funds and burdened by poor trails and unpredictable weather. A dribble of outside capital, especially English money from Shanghai, kept work going. One might wonder how news of Telluride reached halfway around the world to China. Apparently, an Englishman, J.H. Ernest Waters, enthusiastically transmitted it. Waters, a mining engineer, worked in the San Juans off and on from 1877 to 1882, after which he was engaged by the Chinese government as a mining advisor. While so occupied, he organized a syndicate of Englishmen to purchase some Telluride mines, over which he was appointed manager. Other English investors followed his lead; as a result, foreign control here was more significant earlier than it was elsewhere in the San Juans. Several attractions lured the English—the mines' relatively nominal price that mirrored their owners' inability to develop, the district's newness, and the fact that most of the properties showed both gold and silver, not being dependent on only one. Within a few years American investors became interested, including several major stockholders in Durango's smelter. Telluride was fortunate that investment came; without financial assistance, its isolated location might have been its downfall in the 1880s.

Those individuals who invested ran the risk of getting burned. For instance, the buyers of the Golden Chicken Mine at nearby Ophir were plucked of their $75,000 investment in eighteen months. The career of this mine, previously called the Honeycomb and before that the Golden Queen, should have been enough to discourage anyone. All the Golden Chicken laid were lawsuits that reflected ownership and boundary disagreements.

By the end of the decade four mines, all on the same vein, were Telluride's mainstays: the Smuggler, Sheridan, Mendota, and Union. All had been fairly extensively worked and were gaining reputations as solid properties. The Smuggler president, John Porter, reported a yield of $412,000, in the seven years prior to 1889, roughly four-fifths silver, not a large amount by Leadville standards but steady for the San Juans. Already there were indications of an even greater future. Consolidation had joined the Sheridan and Mendota, and the combined operation reduced expenses. High transportation costs were slashed by a tramway, which carried the ore in buckets directly down the mountain from the mine to the concentration mill and brought supplies and intrepid miners back up. A tunnel was being driven into the combined properties through which all future operations would be developed. The benefits of a tunnel for draining, ventilation, and more economical operation proved significant in this case, where it tapped known ore reserves controlled by the company tunneling. In addition, the highly speculative nature of the late 1870s tunnel craze was curbed.

Foreign and/or "outside" control brought with it the mixed blessings of absentee ownership, a reality just being confronted and not yet comprehended. Disillusionment came quickly for some investors. One wrote the Silverton legal firm of Hudson and Slaymaker in 1887 in hopes that it could sell his mining claims. Having no inclination to engage in practical mining and ignorant of the local situation and prospects, this investor cried out for help. Some areas, like Carson, never experienced the heady feeling that came from attracting outside capital. A complaint appeared in the *San Juan*, May 19, 1887, from Chattanooga, northwest of Silverton. The writer was tired of exaggerated reports which undermined undeveloped areas such as his. They only hurt, and he pleaded with investors to come and see for themselves what was available.

Mining expenses stayed high because of the reliance upon freighting, making it all the more remarkable that San Miguel County (basically Telluride, Ophir and a few other scattered mining districts) steadily increased production. Only high-grade ore or concentrates could be shipped and much of the low-grade was simply dumped, awaiting less costly transportation. This district came of age in the 1880s, even without the transportation advantages of Ouray, Red Mountain, or Silverton.[3] Telluride fortunately had mines with enough high-grade ore to encourage development and entice investors; her success stood in stark contrast to the failures of Carson and the La Plata district, which faced similar problems without adequate ore.

The railroad's coming affected the smelter industry as directly as it did mining, the most noticeable impact being the appearance of a regional smelting center in Durango. Shipping ore from the mining districts to a spot where fuel, various types of ore, advanced methods, and skilled specialists were available, rather than trying to spread each of these throughout the San Juans, was certainly logical. Durango's site offered distinct advantages: nearby coal fields; easy travel outlets to the south, north, and west over which fanned roads and rails to tap districts as far away as Telluride, Red Mountain, and Mineral Point; and finally capital, from the backers of the Denver and Rio Grande, which was interested in establishing a smelting works.

Durango proved an excellent choice. Its transportation superiority allowed metallurgists to obtain the necessary mixture of ores to make the lead-based process work economically, something a smelter in Silverton or elsewhere achieved only after high expense and costly delays, if at all. Durango's 6,500 ft elevation offered a more livable environment and, more important, permitted ranching and farming in the surrounding valley, which helped lower the cost of living. All things considered, Durango was more inviting than any mountain location.

The San Juan & New York Mining and Smelting Company, which had secured control of the old Greene works in Silverton, moved to Durango, starting construction of a plant on the west side of the Animas River below the town in 1880. By fall of 1881 the sampling works were ready and the smelters blew in by August 1882. Their completion was timed perfectly with the arrival of the railroad in Silverton. The guiding genius behind the smelter was 31-year-old John Porter; appointed general manager in 1880, he had earlier worked at the Greene smelter in the midseventies, after studying at Columbia and Freiberg mining schools. Not content to work at only one aspect of mining, Porter became involved in nearly all phases. Telluride caught his interest early; Waters, who helped design the Durango works, was probably responsible. Porter invested, wrote reports, and became president of the aforementioned Smuggler Mining Company. A member of the board of directors and vice president of Mears's Silverton Railroad. Porter still found time to dabble in coal, eventually founding the Porter Coal Company. Opportunistic and able, John Porter parlayed the opportunities he discerned into a personal fortune.

Realizing the problems which beset the Silverton works, the company directors tried to retain as much of the operation under their control as possible. They mined their own coal, manufactured their own coke, burned their own charcoal, and maintained scattered mining interests, thus emerging as one of the most important economic forces in the Durango-Silverton area. Porter steadily improved the works, which employed between forty and fifty, and in 1887 smelted $1,012,692 worth of silver, gold, and lead to capture ninth place among the state's smelters. Operation was somewhat seasonal, coinciding with mining; June to October was the busiest period. Reorganized in 1888 as the San Juan Smelting & Mining Company, the Durango smelter illustrated the wisdom of a centralized smelting works' treating ore from a vast region. With the opening of Mears's railroad to Red Mountain and a projected road from Durango to Rico, the nineties looked even more promising for the works.

Porter's success had not been unmarred by jealousy and criticism, particularly for the inability of the smelter to work the Red Mountain copper-based ores profitably. Unsuccessful runs forced these ores to be sent elsewhere, meaning Pueblo or Denver, and resulted in higher shipping costs. A rival smelter appeared in nearby Animas City, which promised to treat just such ore. Porter need not have lost much sleep, because its career was characterized by more shutdowns and repairs than by long runs. Porter met the challenge, and by 1889 the San Juan Smelting & Mining Company had modified its own works, found the necessary ores to use as flux, and was ready to work the Red Mountain ores successfully.[4] That year it prided itself on saving up to 95 percent of the assay value of gold and silver.

Elsewhere around the San Juans, smelting more nearly emulated Animas City's troubles than Durango's good fortune. Mills with poorly chosen sites, badly designed workings, and incompetent management could not keep pace and dropped by the wayside. In contrast to Porter's achievement was the gradual decline, then failure, of the Crooke Brothers in Lake City, which was sold at auction in May 1886. While bond holders forced this action, the company's decline actually duplicated that of the mining district. Troubled throughout these years by ore shortages (their own properties were as much to blame as others), the Crookes shut down several times for short periods, which worried stockholders who saw no dividends coming from investments, and employees who lost jobs and pay. Borrowing more money to satisfy pressing creditors relieved the pressure only temporarily, leaving a millstone the Crookes could ill afford. Even the addition of a concentrator in 1883 could not profitably resolve the woes of low-grade ore and lack of production. The early advantages vanished when the railroad reached first Durango then Silverton, and more prosperous mines elsewhere diverted investors' money that was sorely needed in Hinsdale County for mineral exploration. The Crookes, the best planned and operated rival to Porter, could not equal the advantages of the centralized Durango smelter. There was a lesson here which needed to be more thoroughly studied than it was at the time.

Rico's situation paralleled that of Lake City a decade earlier; isolation (thirty miles and a mountain range from the nearest railroad) was the principal justification for building smelters. Well aware of San Juan history, a correspondent wrote the *Denver Tribune*, September 20, 1882, to argue that the usual objections against local smelters could not be raised against Rico. With plenty of lead ore, nearby coal fields, and local iron and lime deposits, everything necessary for reduction was at hand. Convinced, several firms opened works in the first half of the decade, including the Rico Reduction and Mining Company, which decided that lixiviation was the answer. These Pennsylvania investors found the going rough and were forced to close their works in 1883 because of exhausted ore supply. Not willing to give up, they secured more capital, promising profits of $650 per day on their mines and mill, and added an amalgamation plant. Their hopes were crushed as surely as the ore in their plant. Among other operations, only the Grand View smelter was much more successful; it opened first as a concentrating works, then in 1882 began smelting operations. Its fortunes (producing $148,042 in 1885 and $292,392 in 1887) fluctuated with those of the company's mines, which unfortunately declined, and in 1888 the "old" smelter suffered the indignity of being cannibalized to start a new concentrating and sampling works. The familiar story of not enough ore scuttled Rico's aspirations (something the enthusiastic writer to the *Denver Tribune* had

Winter in the San Juans hampered travel and life in general. One of Otto Mears's Silverton railroad engines passes the Guston mine in Red Mountain.

It was David Swickhimer's faith in the Enterprise mine that brought Rico into its mining heyday. The mine, shown here in June 1890, amply rewarded his faith with silver ore, although Rico never was a second Leadville.

overlooked), abetted by the fact that Rico's ores proved difficult to work. The Rico Reduction Company called them "dry ores" which, because they carried little lead or other fluxing material, were ill adapted to smelting unless mixed with other more amenable ores.

Had Rico's mining fortunes been better perhaps one of these works might have survived. This coquette of a district, however, offered little but flirtation until David Swickhimer found the mining key. One of those captivating individuals who pop into the San Juan story, Swickhimer mined at Rico as early as 1881, only to give up his unprofitable venture because of "wet ground." While trying his hand at several occupations, including running a saloon at Red Mountain City, Swickhimer never lost his interest in Rico. He was convinced that high-grade ore must be at lower levels and purchased half of the Enterprise Mine in late 1886. Though pinched financially and dogged by mounting water troubles, he set about to sink a shaft, racing a neighboring company which held similar convictions about local geology. Buying out less determined partners and holding off creditors, Swickhimer and his miners (many with long overdue wages) hit the pay streak, which ran over 500 ounces of silver to the ton, in October 1887. Good luck rewarded his perseverance—a shaft twenty feet away would have missed the ore body. Swickhimer moved to consolidate and wisely purchased neighboring claims. His Enterprise Mining Company became Rico's most successful.

Thanks to Swickhimer, Rico's fortunes took an upswing in the late 1880s, too late for many investors who had unwisely believed that one or another milling or smelter process could magically turn marginal mines into paying bonanzas. Smelters had a difficult time in the 1880s, partly because, as Theodore Comstock noted, valley smelters had the clear advantage of lower expenses and central locations. When the D&RG opened Durango, Pueblo, and Denver to San Juan ores, no local smelter could compete with respect to reduction cost or mineral percentage recovered. One by one they blew out—Silverton, Lake City, Telluride, and Ouray (needing only climbing ivy to make them picturesque ruins, observed one cynic). They dragged with them angry investors and capital desperately needed for regional development.

Comstock, who wrote a series of articles discussing reduction problems, came to advocate concentration works, which would make it profitable to ship ore to smelters. Practicing what he preached, he established a custom concentrator and sampling works at Silverton, which foundered, according to its owner, because of broken contracts and not enough ore. Comstock gave up and retreated to the academic life at the University of Illinois, having had enough of the vicissitudes of San Juan mining. His failure did not mean the idea was wrong, and the Stoiber brothers (Edward and Gus) opened a Silverton sampling works in 1883. Four years later it could handle 100 tons daily. The Stoibers sampled 6½ percent of each shipment (automatically dividing the ore to improve reliability), thus giving the owner a clear idea of what he could expect to receive from the whole. Maintaining a reputation for accuracy and square dealing, the Stoibers either purchased the ore and shipped it to the smelter or allowed the mine owner to do with it what he would. The brothers dabbled in mining on the side.

Concentrating works proved even more popular, for they allowed lower-grade ore to be profitably mined provided it could withstand the expense of freighting and concentrating. As the decade passed, San Juaners were still reluctant to face the seemingly inevitable reliance they would have to place on outside works. The Gladstone lixiviation plant and one built at Mineral Point in 1887 (despite that abominable winter weather) were examples. Lixiviation had its troubles, especially when imperfect roasting resulted in silver chloride that was poorly suited for chemical reduction. Less important, the process had difficulty producing silver bullion free from lead. Although works like these at Gladstone closed, the process looked good in theory and San Juaners were reluctant to give up. A special correspondent to the *Engineering and Mining Journal*, August 4, 1888, lamented the many "so-called mills" built in days gone by which stood as solitary monuments of "ignorance and folly."[5]

One form of reduction, the stamp mill (ideally combined with concentrating machinery), made a comeback, particularly in San Juan County, but also in some of the smaller gold-producing areas. Rather than evidence of disillusionment with other reduction methods, it showed the upswing of interest in gold and the continuing search for relief from the low-grade ore problems. San Juan County, for instance, even in the gold fever days of the early seventies, had never mined over $20,000 per year of the metal. As the mideighties passed, this figure climbed and reached $394,000 in 1889, temporarily jumping past silver. Less startling was San Miguel's production, which leveled off at a ratio of 6-4 in favor of silver. Possibly the renewed interest in gold was merely a by-product of increased silver mining, but more likely it hinted of something else. Alarmed San Juaners watched as the average per ounce for silver slipped from $1.15 (1880) to $.94 (1889); what four ounces used to return now took five. As early as 1884 Horatio Burchard, in his report as director of the mint, remarked that the decline in price of silver and lead had caused many owners to hold back their ore in hopes of rising prices. He also observed that some low-grade ore bodies could not be profitably worked. Silver was struggling—not even the optimistic could deny it—just when local silver mining had finally turned the corner.

In retrospect, the factors behind this alarming decline are easily identified: swelling production and decreasing utilization. The San Juans contributed their share to the increase, small though it was compared to Aspen over in Pitkin County or Leadville. The international decline reflected the decrease in use of silver coinage, as countries shifted to the gold standard rather than the bimetal (silver and gold) or silver alone. The other market source, industrial and commercial use, could not absorb the surplus by itself. Generally outspoken on the issue, seldom completely understanding all its ramifications, and never lacking emotion, San Juaners grew steadily more vociferous on the silver question. Lively discussions on cold winter evenings centered upon causes, effects, and cures; they produced much verbal heat but shed little light. A political candidate's stand on silver could become the touchstone for local success. But in mining nothing much could be done except to ride out the depressed conditions, or switch metals. Unlike a farm, a mine could not stand idle, and storing its ore was not feasible or practical because of expenses and depreciation. There remained the one slim possibility of reducing costs, which would be hard to achieve once the cost of living had stabilized and the railroads had standardized their rates. From the owners' view the point to attack was the miners' wages, a questionable solution.

This concept led to the first outbreak of labor trouble in the San Juans. Telluride had problems in 1883 and again in 1886. The first incident involved the whole camp and stopped all mining, but resulted in success for the miners; the second confrontation collapsed when some miners accepted a reduction to $3.50 a day, leaving the Sheridan miners alone in their defiance. They had no choice but to concede, especially when Telluride paid fifty cents more per day than neighboring Ouray, an indication of higher living costs and isolation. Interestingly, management promised to restore the cut as soon as silver reached $1.02. Over in Silverton in 1885, John Porter found himself in similar trouble when he announced a fifty-cent wage cut at the Aspen Mine to two dollars per day. Justifying his action on the basis of the continued low price of lead and silver, Silverton's "higher" wages, and the mine's isolation which kept freighting expenses high, Porter faced immediate opposition not only from his own men but also from those of Red Mountain, Mineral Point, Animas Forks, and Silverton. The reduction was revoked when no miners could be found to work at the new scale. Underlying bitterness flared when a foreman of the National Belle Mine at Red Mountain, who favored reduction, received "two sound thumpings" within one week from irate miners.

Not unexpectedly the miners turned to unions for protection. The Knights of Labor, open to all workers, had assemblies at Silverton and Red Mountain that were dominated by miners. Discussion forums primarily, they also were active socially, sponsoring dances and marching in July Fourth parades. An unidentified miners' union was involved at Telluride during the 1883 strike, and the Knights applied pressure against the Aspen Mine.[6] Battle lines between *(right) Miners, such as Silverton's Ben Harwood, were the backbone of the nineteenth century San Juans. Harwood holds a single jack hammer and drill and has a pick next to him, although he was obviously sitting in a photographer's studio.*

(below) Just above Telluride was Pandora, where the mills and smelters were built. The Pandora mill was one of the early ones. In this September 1886 scene, mules are hauling mining timbers up to some mine.

management and labor tightened, each side warily watching the other by decade's end. Absentee ownership whipped up further resentments; the wind was being sowed.

Mining innovations also lowered costs, generally with less friction. Tramways, bucket-carrying cables supported by towers, stretched from mine to mill or railroad siding, offering unmistakable advantages in year-round operation. With gravity's assistance, they overcame geographic obstructions and lowered costs; full buckets coming down pulled empty ones back up. The freighter and burro could not compete. Trams could be constructed almost anywhere— over cliffs, up mountains, across streams—and, in theory, were cheaper to maintain than roads; a snow slide, however, might carry away a couple of towers and alter that advantage. Versatile, they carried ore out and brought supplies in, and venturesome miners traveled relatively quickly. Swaying to and fro in a bucket, numbed by the cold wind, many a miner wondered if it would have been wiser to walk. Yet it beat hauling coal by wagon part way, then packing it on burros to reach the mine, and reversing the procedure for ore coming out. The Little Annie, near Summitville, used a tram to transport ore from mine to mill, and a 1,500-ft one was built by the Grand View Company at Rico. The Cosmopolitan Mining Company (Ouray) enticed investors by describing how its tram allowed winter work, meaning more profits.

Already comments were being heard about electricity and its possible uses in mining. Too new to make much of an impact, electricity remained unnoticed as the 1880s closed. Year-round operations, at least on the larger properties, became relatively standard, solidifying the trend noticed in the late seventies. Nothing changed with miners, however, who had months of fun to catch up with and money to satisfy their whims when winter at long last released them from their snow-locked boarding houses. Patenting claims to secure titles also became more common. After a claim was filed, assessment work amounting to $100 had to be performed each year or the claim was forfeited. Most San Juan claims got no further; however, if the owner thought the property especially valuable after an outlay of at least $500, he could apply for a patent. This generally required a lawyer's help and involved a survey, patent notice publication in the nearest paper, sworn statement of ownership, and finally, if no adverse claim was filed, paying for the property at five dollars an acre, a real bargain. Only then was the claim safe; it was a bothersome procedure in the short run, but much cheaper over the long run.

Vigorous promotion of the area continued, especially in the early 1880s. For example, at Denver's National Mining and Industrial Exposition, 1882-84, San Juan mines and counties were well represented. Those two war horses, William Weston and Theodore Comstock, occasionally crossed pens in their enthusiasm, ruffling each other and their followers. Comstock continued to favor San Juan, and Weston, Ouray County, and never would the twain meet. Comstock's departure removed one of the prime promoters but others rushed to fill the gap. Nothing discouraged them, not even the fact that the San Juans seemed destined always to be second rate. "Booms are bubbles consisting of tailings of emigration, which is a detriment to any country," wrote a Telluride booster, a sentiment Weston agreed with, almost:

> Another year has passed, and I rejoice to say, without a 'boom' in San Juan: for in most cases I find that a 'boom' in any new camp means undue excitement, fathered by prospectors, nursed by assays of rich picked specimens, fostered by fictitious local newspaper reports, and brought to an unhealthy maturity by a rush of deluded people.

By the end of the eighties direct contact with potential buyers was replacing the casting of tempting bait to the general public. The San Juans were nearly well enough recognized as a mining area not to need the type of promotion typified by earlier efforts.[7]

Particularly encouraging for the development of hardrock mining, smelting, and the region's general prosperity was the growth of coal mining, spearheaded by La Plata County.

The Stoiber works at Silverton were busy on this day. The mules have brought ore to be sampled for mineral content; here was where the fate of a mine was decided. The Stoiber brothers eventually got into both ends of mining.

Predictions as far back as 1878 pointed to coal's bright future, and several parties were then engaged in what was termed "quarrying" the coal for use at Animas City. Because the seams were open on the surface, it was not surprising that locals whacked away at them. From such beginnings, seven mines produced 33,280 tons in 1889, most of it from the San Juan Mine. Though far behind the eastern slope's big coal producers, it was a good beginning and augured well for the years ahead. Coke ovens near Durango further diversified the product and helped to provide more jobs. The smelter's need for coke could be met locally; it was a happy coincidence that the best coking coal was near the San Juan works. Other coal deposits were reported in San Miguel, Ouray, and Dolores counties, although only the last showed any significant production. The Rico smelters owned or leased these mines primarily to insure a fuel supply and because they coked well. Local coal would find a ready market if its quality and volume were equal to that being offered by outside competition; only time would tell how far the coal industry could go to match this challenge.[8]

Another encouraging sign was the appearance of trained mine managers and mining engineers. Though training and experience did not guarantee success, they did hold out more promise than faith, hope, and a lucky guess. Thomas Comstock sagely observed, nevertheless, that all the combined power of envy and prejudice opposed the advantage of training and experience.

One of those trained individuals was mining engineer Eben Olcott, who worked at Lake City and Silverton in 1879-81, spending most of his time at the North Star Mine above the latter. Olcott's letters provide a peek into his world. The North Star bunk house, as he described it, possessed a character all its own, with hams, bags of coffee and sugar, shovels and picks, the stove, dining table, and double bunks three tiers high vying with twenty men for the available space in the single room . "When men's boots got piled up round the stove to dry and every man begins to smoke a vile pipe. I will stop, you can imagine the rest." Of the stove, he moaned, "I am alternately freezing one half or other for the stove has no effect upon that side turned away from it." The weather and the countryside continually caught his attention and he even had time to go skiing. Writing his wife very little about mining, apparently not caring to bore her, he once let down his guard to accuse the company of wasting money, largely because they did not heed his suggestions. In the end, Olcott moved on. "I hate to change but it seems necessary for a mining engineer to do so pretty often. In one sense this is very valuable and desirable for it enlarges our scope."[9]

Mining in the eighties remained much more like that of the seventies for most owners, highly speculative and exciting. Ernest Ingersoll put it this way for the readers of the *Harper's Monthly* in April 1882: "Everybody looks forward. Each proposed to do this and that, and to be happy, 'when I sell my mine.' Perhaps this delicious uncertainty is part of the fun." In 1886 Otto Mears and several others became interested in the Buckeye Mine above Silverton in Ice Lake Basin. After carefully examining it, they purchased what they considered a bonanza, "the biggest mine in the San Juan." Over the next two years expenses ran high for wages, supplies, coal, freighting, equipment, and ore sampling. In February 1888, a "big" snow slide raced past the mine, scaring some of the miners half to death, and they scurried off the mountain. Throughout that year, each week seemed to improve prospects; hopes climbed, only to crash in 1889 when the mine shut down. No reason was given, except the cryptic note on the last four samples sent from the Buckeye—two showed nil, one a trace, and one a small amount of silver.[10] The "biggest mine" had come a cropper.

The San Juans ended the decade on a much more secure mining base than they had started it. Despite the glitter of Leadville and Aspen, investors were finally coming in and transportation had certainly improved. Though not yet a byword in American mining circles, the San Juan mining districts had shown their staying power and diversity in the eighties. The time of waiting had nearly passed.

[1] Sources on San Juan railroading: Robert G. Athearn, *Rebel of the Rockies* (New Haven: Yale University Press, 1962), pages 14, 43-44, 104-105, 134, and 160. J. Davis, *Narrow Gauge Railroads* (San Francisco: George Spaulding, 1880), pages 8-11. Robert F. Weitbrec Papers, Colorado Historical Society. Josie Crum, *Three Little Lines* (Durango: Durango-News, 1960), pages 1-5. *Engineering and Mining Journal*, December 30, 1882, page 350; February 17, 1883, pages 91-92; August 4, page 91; September 22, 1888, page 244; and July 6, 1889, page 13. See also Robert Sloan and Carl Skowronski, *The Rainbow Route* (Denver: Sundance, 1975). For mine production see Henderson, *Mining in Colorado* and H. Burchard, *Report of the Director of the Mint* (Washington: Government Printing Office, 1884), page 235.

[2] Sources for Summitville and Lake City mining: *Engineering and Mining Journal*, 1880-89; *Lake City Mining Register*, 1883; E. S. Larsen, "The Economic Geology of Carson Camp, Hinsdale County, Colorado," *Contributions . . . Geology* (Washington: Government Printing Office, 1911), pages 30-38; Henderson, *Mining in Colorado*. Scattered issues of the *Rocky Mountain News*, *San Juan Prospector*, and *Silver World*, 1880s; Burchard, *Report* (1883), page 329.

[3] Sources for Red Mountain and Telluride: *Engineering and Mining Journal*, 1880-89. *Financial and Mining Record* (New York), 1889. Charles Leonard, "In the Red Mountain District," *Colorado Magazine*, (April 1956), page 149; *Red Mountain Pilot*, 1883; *Red Mountain Review*, 1883. *Denver Tribune*, 1882-83. J.A Porter. "Report on the Smuggler Mine," (February 15, 1890), Hague Collection, Henry E. Huntington Library. *Mendota Mining Company* (Shanghai: North-China Herald, 1890), pages 2 and 8-11. *Silverton Weekly Miner*, January 4, 1890. R. Hopkins to N. Slaymaker, April 15, 1887, Hudson and Slaymaker Collection, San Juan County Historical Society.

[4] *Denver Tribune*, September 14, 1882. *La Plata Miner*, September 11, 1886. San Juan & New York Mining & Smelting Company. Ledger, 1885, First National Bank of Durango. *Reorganization of the San Juan & New York Mining & Smelting Company* (New York: np, 1888), page 10. *The San Juan Smelting & Mining Company*, 1888-89 (1890c), pages 1-7. T.A. Rickard, *A History of American Min-*

ing (New York: McGraw-Hill, 1932), page 122. *Engineering and Mining Journal* 1880-89 and September 26, 1891, page 239. *Red Mountain Review*, May 12, 1883. *Daily Herald* (Durango), August 2 and September 9, 1882. *The Idea* (Durango), September 27 and October 19, 1884.

⁵ Sources on San Juan smelting. Lake City: *Engineering and Mining Journal*, 1880-86. Rico: *Dolores News*, 1882; *Denver Tribune*, September 20, 1884; *Engineering and Mining Journal*, 1882-89; *Prospectus . . . Rico Reduction and Mining Company of Rico, Colorado* (Tyrone, Penn: C. Jones, 1883), pages 3-5, 9, and 11-17; Rickard, *American Mining*, pages 135-139; Frederick Ransome, "The Ore Deposits of Rico . . . ," *Twenty-Second Annual Report of the United States Geological Survey* (Washington: Government Printing Office, 1901), volume 2, pages 308-309. *Lake City Mining Register*, 1882; *San Juan Democrat, 1888. The San Juan*, 1887. *Solid Muldoon*, 1880. *Denver Tribune*, September 21, 1882. *Report of the Director of the Mint*, 1881-89. C. Stetefeldt, "On the Process . . ., *"Report of the Director of the Mint* (Washington: Government Printing Office, 1884), pages 743-745. For general background see James E. Fell, *Ores to Metals: The Rocky Mountain Smelting Industry* (Lincoln: University of Nebraska, 1979).

⁶ *The San Juan*, July 14, 1887. *La Plata Miner*, November 29, 1884 and January 17-February 14, 1885. *Solid Muldoon*, November 9 and 30, 1883. *Animas Forks Pioneer*, February 7, 1885. *Tribune Republican*, August 7, 1886. *Lake City Mining Register*, December 7, 1883. *Engineering and Mining Journal*, 1883 and 1885. Colorado Bureau of Labor, *First Biennial Report . . .* (Denver: Collier and Cleaveland, 1888), pages 118-120.

⁷ *Engineering and Mining Journal*, January 8, 1881, page 22. *San Miguel Journal*, quoted in *Rocky Mountain News*, February 15, 1884. See also the *Solid Muldoon* and other local papers for various promotional material.

⁸ *Rocky Mountain News*, April 24, 1878. *The Southwest* (Durango), February 9 and 16, 1884. *Financial and Mining Record* (New York), June 6, 1885, page 354. *The Idea*, December 26, 1885. Burchard, *Report of the Director of the Mint* (for 1881), page 420. *Mineral Resources of the United States* (Washington: Government Printing Office, 1885), pages 30 and 31-34. See also Reports of the Colorado State Inspector of Coal Mines, 1884-91.

⁹ Olcott to sister, November 4, 1880 and July 29, 1881. Olcott to wife, June 16, September 14, October 12 and 17, 1880, and February 6, 1881, Eben E. Olcott Collection, University of Wyoming.

¹⁰ Journals, 1886-89, Weitbrec papers. Ernest Ingersoll, "Silver San Juan," *Harper's Monthly* (April 1882), page 703.

The
San Juaner

The San Juans were dotted with log cabins such as this one that William H. Jackson saw in La Plata Canyon in 1875. Jackson called its owner a "hermit-like miner," a fit description for many prospectors who wandered into the region.

P hotographers came to the San Juans as early as the mid-1870s, the most noted one William Henry Jackson, who toured the region in 1874 and 1875. They preserved almost every aspect of a San Juaner's life.

Each of these photographs has a story to tell. There is a warmth here that the cold pages of history cannot convey, an understanding that words alone cannot give. These photographs reveal a world long removed from the present, and they do so without romanticizing it or falling into the trap of nostalgia. It is as if a curtain parts for a moment to allow a glimpse of history stopped by a camera.

The real West, it has been said, is not where and when, but who. The who of this essay is the San Juaner, that otherwise unidentified individual who is the heart of this story. A writer in "Ecclesiasticus" observed, "Some there are who have left a name behind them to be commemorated in story. And some there be, which have no memorial." This chapter is a tribute to the latter.

(top) Gladstone never matured beyond the status of a mining camp. Main Street was nothing fancy, yet these well-dressed, turn-of-the-century folks seem to be heading for some sort of celebration at or near the saloon.

(bottom) Ironton was at its peak in this 1890s view. The Silverton Railroad came by here, and mule trains moved goods from the camp to the more isolated sites. A bicycle craze hit the San Juans in the nineties. At its best Main Street was rough and dusty; mud was characteristic in season.

(top) Animas Forks' main street pointed the way to Silverton far below in the valley. Signs straddling the boardwalk were a common way to advertise. In 1882 the business district included general merchants, saloons, a livery stable, blacksmith, hotel, drug store, and meat market. False fronts were typical architecture.

(bottom) A quarter of a century can make a big difference in a mining region. This couple off for the mountains takes bed springs and a mattress. Telluride also shows its maturity with board sidewalks, fireplugs, and homes that appear comfortable.

(top) San Juan communities were vulnerable to fire, even when they had a fire company and a water system. If the mines and local economy continued to be prosperous, rebuilding might produce a better planned and constructed settlement. In Red Mountain, this August 14, 1892 fire destroyed too much and there was too little incentive to rebuild. The "Sky City," as it was nicknamed, hung on, only a shadow of its former self.

(bottom) Boarding and rooming houses were common in the masculine mining world. They were quite often a family-run business, as this one seems to have been in Telluride.

(top) Frank Hartman (left) and Charles Jones (second from left) purchased the News in April 1880, promising in their first issue to make it "in every sense a live paper." Rico had six papers during the 1880s, with a high mortality rate. The fellow on the right is the printer's devil, an invaluable member of the newspaper "team."

(bottom) A stable or two was a must for providing a carriage to take one's best girl on a whirl or to transport a prospective investor or purchaser to a mine. This Ouray owner was obviously proud of his horses and buggies. The Ouray Times next door published from 1877 to 1886.

Ruth Gregory, Ouray

(top) A mining camp would not have had an opera house and permanent dentist, such as these in Ouray. Wright's Opera House is a magnificent example of a cast iron front, not the more common wooden false front. This particular evening the hall was going to be used for a dance.

(bottom) Every San Juan community had to have a hotel, usually the first sight for visitors as they stepped off the stage. Many were fancier than Howardsville's Hotel Galena. Next door is that familiar fixture of the mining camp, the general store.

Pioneering in the San Juans

(top) Two major holidays were celebrated in the camps—Christmas-New Year and July Fourth. Baseball games, hard-rock drilling contests, and hose races were favorite events on the Fourth. Despite the Spanish-American War, Telluride was in a holiday spirit on July 4, 1898.

(bottom) Tennis at over 11,000 ft. soon took the wind out of the players. The Tomboy Mine provided the court and also indoor bowling. Down at Telluride, baseball, basketball, and other activities awaited the sports minded in post-1900 booming Telluride.

(top) A horse-drawn school bus parked in front of Telluride's high school. Telluride's school system was one of the best in the San Juans, reflecting the town's prosperity after the conclusion of the labor unrest in the early 1900s. Most camps had to be satisfied with the "traditional" one- or two-room school.

(bottom) Much of the entertainment and relaxation was home grown, such as this excursion party near the Camp Bird mine. Hunting, fishing, berry picking, sleigh rides, and skiing occupied the San Juaners at various times during the year. Someone thought the occasion called for more than the casual clothes normally worn on a picnic in the mountains.

(top) Lawlessness was not a thing to be condoned, nor did a man "feel naked without his gun." Few San Juaners carried them; a lawful image helped attract settlers and investors. Durango turned out for its only legal hanging in June 1882. Apparently some parents thought it was a good object lesson for their children.

(bottom) San Juaners were fanatical in their loyalty to William Jennings Bryan, shown here in front of the New Sheridan Hotel in Telluride. Otherwise, they were maverick in their political allegiance but always loyal to their own and the San Juans' interests, as they interpreted them.

(top) Circuit riding ministers came
early. The first sermon was preached on
July 5, 1874 to a motley crew at Howardsville. Those early congregations seldom
had the family flavor of this 1898 gathering
at the Baptist Church in Lake City.

(bottom) Death was a constant companion
for all those who dared the San Juans,
whether on the surface or underground.
This party is bringing home two victims of
a snowslide in Hinsdale County. Slides
claimed more victims than any other natural disaster in these mountains and continue to this day to trap the unwary.

By the mid-1880s refinements had come to the San Juans, as this Silverton sitting room attests. Obviously no bachelor miner's cabin, this represents Victorian elegance as interpreted by a prosperous San Juaner.

The 1880s visitor to the San Juan mining communities observed a mixed bag of styles and varying degrees of maturity. Some camps gloried in newness, in the best of western tradition; others, wearing at least a decade of age, tried their best to hide the wrinkles and display their maturity. Mining engineer Eben Olcott wrote his sister that "very much of a feeling of civilization pervades" in Lake City. A season spent in the mines might have influenced his reaction. Yet Capitol City, with its brick construction, and Animas Forks, with its three-story building—both small camps

"Poor Oats are Hard to Sell"

—strove for that same mature appearance that constituted such a large part of the San Juaners' concept of civilization. These little settlements even acquired the promotional trappings of late nineteenth-century America. Poughkeepsie was the "biggest little mining camp of the San Juan country," and Telluride "The Golden Gem of the Silvery San Juan." They readily accepted progressive innovations such as the telephone, partly because of convenience and partly because of the image they projected.

Ernest Ingersoll, who toured the Durango-Silverton-Ophir area, was moved to comment that "there is no Topsy growth at all; rather a Minerva-like maturity from the start." A characteristic of these mountain "villages," he concluded, was that they sprang full size into both existence and dignity.[1] Well, yes and no, but locals could not help but be proud of so favorable an impression. What Ingersoll portrayed as pioneer settlements had actually passed through that period; as other tourists before and since, he approached them with a preconceived idea of what he would see. With such a prejudgment, Silverton appeared well planned. Durango, from its inception, was no ordinary San Juan community—the Denver and Rio Grande and its organizers saw to that. Planning and substantial development marked its early career; perhaps it could truly be said to be "Minerva-like." To call Ophir "Minerva-like," with its log cabins strung out beside a dusty road, was to fall victim to one's preconceptions.

Ingersoll no longer traveled in a frontier land, no more strikingly demonstrated than by the plush accommodations offered by the Denver and Rio Grande. The "very rough" bridgeless trails of George Darley's day had become rough wagon roads stretching to the far corners of the San Juans. Over them came a better variety of food and a larger selection of general merchandise. Fastidious travelers might still describe the accommodations as primitive, even in the larger communities.

Urban growth, which so amazed Ingersoll, continued unabated into the eighties, every new mining district convinced it needed a "metropolis." Durango, by far the most distinctive newcomer, never had been a mining camp in the truest sense. As mentioned, it was a smelter town, a transportation center, and a supply point for the nearby mining districts. Durango combined the natural advantages of location, climate, and resources with its transportation advantages and aggressive business and political leadership (gaining the county seat within its first year and a half) to become by decade's end the San Juans' largest town.

Within the mining districts the most impressive urbanization unfolded in and around the Red Mountain mines. Here Guston, Ironton, Red Mountain, and Red Mountain City came into being, all within five miles of each other. The last two were never compatible, feuding and fighting throughout their brief rivalry. Only a mile and a mountain top apart, one in Ouray and the other in San Juan County, they had little favorable to say about each other. Red Mountain City finally lost its newspaper and its name to its better-located rival and became Congress, merely a stop on Mears's railroad.

In the older districts mining settlements less frequently broke new ground, a relocation or consolidation being more common. Ames, near Ophir, was typical. A promising, "wide-awake village" in the early 1880s, Ames quickly slipped, as a reporter quaintly observed, "to the shady side of life" when its smelter failed. Unfortunate Parrott City lost its county seat designation to Durango and its reason for existence to the town of La Plata farther up the canyon near the mines. La Plata, an "enterprising little place," never grew beyond that, nor did its contemporaries, Middleton, between Eureka and Howardsville, or White Cross over in Burrows Park. An 1885 count found thirty-three communities scattered throughout the San Juans.

In the eighties the larger mining towns steadily came to dominate, as the pretense and aspirations of the smaller camps slowly faded away. The difference between a mining "camp" and a mining "town" was unclearly defined and difficult even for the people of that day to explain concisely. The distinction was based partly on population, the size and type of business district, wealth of the community and surrounding mines, architecture (brick and stone, multistoried structures), and life styles. This last factor had to be seen and sensed rather than vicariously imagined decades later. The mining towns possessed a certain spirit, a certain attitude, which the visitor could feel; a mining camp, on the other hand, aspired to this essence, only to find that it rang untrue in the smaller setting. The mining town category included Ouray, Silverton, Rico, and Lake City, with Telluride knocking on the door by the late eighties. The Ouray-Silverton-Telluride cluster was on the rise; both Lake City and Rico were having their troubles.

The San Juaners crowded into these communities and lived at the surrounding mines. Ordinary individuals, holding no particular claim to fame, they passed their days unrecorded and scarcely appreciated, working for a living, hoping that someday fortune might smile on them. Their fate evokes regret, because it was they who provided the sweat and much of the substance of the San Juan story. Twice in the 1880s, census takers visited them to record a statistical portrait. From this they compositely materialize.[2]

They came from all over the world and the United States, drawn by the magnet of mining. Native-born Americans predominated, their numbers swelled by the rising percentage of Colorado-born children. Regionally, the East and Midwest sent the most. The largest foreign element emigrated from the British Isles, trailed at considerable distance by the German states. The San Juaners were overwhelmingly of white, northern European stock, possessing a common cultural and linguistic heritage, English.

The odds were better than ever that a random sample would be male, single, in his early thirties, and American born, as were both his parents. Equally good odds could be given that he was a miner and had been in other districts before coming to the San Juans. With the addition of other occupations related directly to mining (for example, mining engineer, assayer, and most of the blacksmiths), the dominance of mining in the economy was firmly established.

The unsung heroes were the laborers. Found in all the unglamorous jobs—working at the surface of the mines, building roads, pushing wheelbarrows at smelters, hauling freight at depots—they did hundreds of other mundane but crucial tasks. Common labor was the starting point for many a young man, shown by the fact that this category had the lowest average age of the working groups (midtwenties); most were single and American. The Germans and Irish made inroads though, and the latter and their English cousins were also found underground as miners. Not so the Germans, who preferred working where they could see daylight, as in blacksmithing.

On Main Street waited the merchants, ready to relieve any and all of their wages. The ever-present general merchant was one of the oldest of the group sampled, at slightly over 37 years, and he was generally married. Americans predominated, with English and Germans, as expected, leading the immigrant contingent. His age, marital status, and size of investment made it easy to understand why he worked so diligently to create a lasting settlement.

His neighbor, the saloon keeper, was nearly as old, but overwhelmingly single. The members of this group came from all over Europe and split the occupation just about evenly with Americans. Unaccountably, New Yorkers seemed more likely to tend bar in the San Juans than any other state representatives. Unlike saloon keepers, hotel operators were married, probably because this was a family occupation in which both husband and wife were needed.

The professional class tended to congregate in the larger communities, as mentioned earlier. Doctors, lawyers, engineers, ministers, and dentists—the majority of this group were American born, slightly older, and leaned more toward matrimony than the miners. Their average age ranged in the thirties, the attorneys being the oldest, at barely over 40. New York contributed the most doctors in an American-dominated profession. Ouray had something unique, a widowed Kentucky lady who gave her occupation as "doctress."

The exceptions to the above masculinity in the professional classes were the teachers, who quite likely were American, female, young, and single. The few male "professors" received better salaries in this day before equal pay for equal work. Far outnumbered overall, women yielded to none as an influence in the social, cultural, and religious life of their communities. Generally married (very few slipped past their midtwenties before some lonely bachelor proposed), they came from throughout the United States. The percentage of American born was much higher than among the men. Being a homemaker and mother occupied the majority, but a few competed successfully in the job market. Besides teachers and hotel operators, they worked as waitresses, cooks, seamstresses, domestics, and boarding house and restaurant managers. One was a telegraph operator. Silverton had a female barber and Durango and Rico had women newspaper editors.

These then were the San Juaners. Equally fascinating would have been a look at the other businessmen along Main Street; not enough representatives existed, however, from which to cull a composite picture.

The census takers found few blacks or Orientals, far fewer than local sources indicated lived there. Lake City, Ouray, Rico, Red Mountain, and Rose's Cabin had, at one time or another, black miners and prospectors, and others worked as barbers and waiters and ran a dairy near Lake City. How many Chinese ventured into the San Juans cannot be determined. They opened those traditional "washee" establishments and restaurants, and owned and operated a mine on Red Mountain in 1883. At least one anti-Chinese riot erupted. In 1882 a mob of masked men beat up and then drove eight Chinese out of Rico, ransacking their quarters in the process. The *Dolores News* was appalled, as was the town council, which sent a wagon to bring them back. The matter was then conveniently dropped in the press and the aftermath was lost to history.[3]

Immigrants of all types gravitated or were pushed toward certain occupations in a much higher ratio than their overall percentage should have merited. They were found as blacksmiths, miners, saloon keepers, general merchants, and common laborers, doing jobs that usually required a great deal of physical stamina. Only rarely did they break into categories defined as white collar. Americans also dominated in the slowly growing farming and ranching areas on the San Juan fringes, despite heavy foreign involvement in these occupations in the Great Plains and Midwest.

The market for agricultural and ranching products grew with mining; more than just food for human consumption was involved. The need for grain to feed the horses, oxen, burros, and mules that kept the San Juans moving became acute. William Pabor, an early Colorado agricultural booster, was enraptured by the potential of La Plata County's valley lands. He con-

Even in the early 1880s, Telluride was showing signs of the prominence that would come in a decade when it became the San Juan's greatest mining town. Mary Mott, who lived there, particularly remembered the ladies' horseback club and its rides on beautiful fall days. She also recalled the Cornish miners and how they loved to sing.

ceded, however, that even if they were all occupied and cropped to utmost capacity, the supply would simply not equal the demand, in "consequence of the rapidly increasing population of the mining districts." The need was met by farms that far exceeded the family-subsistence level. Peter Archdekin owned 1, 000 fruit trees, as well as a dairy herd, and raised everything that could grow "at this altitude" on his home ranch six miles from Durango. He supplied Silverton and Durango with vegetables and small fruits and branched out with his Home Ranch Meat Market. Durango was not alone; Ouray prided itself on being the only "first class mining camp" in the state that "is self sustaining in the agricultural line," while a Telluride editor estimated 12,000 head of cattle were on the summer range west of town.

Profits did not automatically accrue to the rancher and farmer, who still needed to acquaint themselves with the growing season, rainfall patterns, and plants that would mature rapidly. As many a would-be gardener found out, a high-country killing frost could hit as late as June. The arrival of the Denver and Rio Grande proved only a partial blessing to agriculture. It brought lower freight costs, while also opening the San Juans to eastern Colorado and Great Plains competition. The newcomers stayed and their numbers increased, shown clearly by the establishment of purely agricultural settlements such as Cortez (in 1886). The farmer and the miner generally got along famously; they needed and profited from each other.[4]

Customers were becoming more selective and would not readily purchase a second-rate product, as freighter and supplier Dave Wood found in 1887. The oats in a grain shipment, complained the purchasing merchant, were "pretty hard, a good share being colored, musty and dirty." He pleaded for good ones, as "poor oats are hard to sell." Ten years before, customers would have been satisfied merely to have received any grain shipment, with no complaints about quality. The improved transportation network and standard of living encouraged San Juaners to expect and demand better. To select Dave Wood for the above illustration was perhaps unfair, because he owned one of the largest and certainly one of the finest freighting outfits in the San Juans, working out of Montrose. Even after the railroad came, the freighter continued to be essential in keeping the local economy in gear. For example, in 1885, Wood hauled 2,376,438 pounds of freight to Ouray and hauled out over three million pounds of ore, nearly half of his grand total that year of 11,965,035 pounds.

His correspondence provides a fascinating glimpse into just how refined the San Juans were becoming. The president of Telluride's San Miguel bank wanted him to ship carefully an ore cabinet case, while an impatient lady in the same town wondered where her case of wine had gone. A merchant at Red Mountain inquired what the rates were for shipping powder; if reasonable enough, he intended to ship via Montrose rather than Silverton. Credit caused Wood grief; an Ames general merchant apologized for not paying his bill because of the large accounts outstanding, which he had been unable to collect. On the side, Wood himself indulged in merchandising. A Denver brewer advised him he could handle its beer "for your mountain trade at wholesale" and offered him a 5 percent discount on the $10 price per barrel. He also sold horses and teams; one purchaser returned "Shoestring," who turned out to be "hell in the stable," trying to climb out of his stall and causing a "general ruckus."

One of Wood's rivals was the Mears Transportation Company, part of Otto Mears's little empire. Mears, who had done a great deal for the San Juans in the 1870s, finished his toll-road system, turned to railroads, and emerged a hard-headed businessman. Too hard headed for some people, who complained about the condition of his roads and "outrageous" charges. In Mears's defense, it should be emphasized that just maintaining and operating his roads was a continual headache; he had to employ toll-gate keepers and laborers to "shovel and pick" to remove snow and repair the road and deal with a large number of other problems. An opportunist at heart, Mears steadily enlarged his initial investment and enthusiasm into a district-wide business empire and entrepreneur status for himself.[5]

Business districts changed along with the rest of the community. The new camps looked much like their predecessors of the previous decade. Red Mountain offered a saloon, hotel, general merchant, and meat market. The older ones, such as Animas Forks, had the same things, plus a blacksmith, laundry, drug store, barber, livery stable, and restaurant. A mining town had all these and more. By 1888, Silverton could boast of such specialty shops as clothing, boots and shoes, bakery, hardware, furniture, and fresh fruits. As had Lake City's back in the seventies, Silverton's business diversity attracted more customers and profits, hence more businesses. Each settlement suffered recessions, which crippled the weaker enterprises, but on the next upward swing, other optimistic individuals came in to replace their fallen brethren. By carefully studying these changes in a community's business, the fortunes of the district may be chronicled.

National advertisements and mail order companies were making increasing inroads on the San Juan scene. Rico's *Dolores News*, in February 1882, carried not only local advertising but also ads from Durango and Montgomery Ward, and others that exalted numerous patent medicines which were virtually certain to cure whatever ailed one. Four years later, the *Animas Forks Pioneer* displayed among its advertisements those for the Chicago Cottage organ, the Shorthand Institute of Louisville, Kentucky (accessible by correspondence), Mexican Mustang liniment (a family medicine), Tower's fish brand slicker (the best waterproof coat), Dr. J. Stephen of Lebanon, Ohio, who promised to cure the opium, or morphine, habit in ten to twenty days, and Clingman's tobacco remedies, the sure cure for itching piles, sprains, ulcers, sore throat, dog bites, and corns. Local merchants responded to this onslaught by reducing prices, improving displays, and arguing that people should trade locally. Regardless, the mail order business was there to stay, taking a sizable bite of the general merchant's trade.

It did not take much to diminish profits. F.C. Sherwin ran a general store at Silverton, and a random selection of his day's total business from January through May 1889 found a low of $6.33 and a high of $48.35. Once, for example, the purchase of a $25 suit of clothes and a $4 hat in the same order represented 75 percent of the day's business.

Looking at Main Street more closely, it may be observed that the towns cornered the banks, those significant symbols of local business growth. Bank transactions and deposits mounted steadily, resulting in a larger geographical representation of clients. Problems arose because the tellers no longer knew the depositors personally. As long as someone could sign his name, there was no impediment, but a minority were unable to do so. Durango's First National Bank resolved this problem by describing some identifiable feature, for instance, "lost part of forefinger of right hand," "colored very black about 5 ft 4 in." and "37 years old, moustache, lives in Durango." A warning to new employees about one depositor read, "when money is drawn the lady to accompany gentleman."

All the towns also became county seats (court sessions inspired an economic boom; a stagecoach full of people left Animas Forks for Silverton one June day in 1882 to appear as jury or witnesses), and most acquired a coterie of lawyers. They did not spend all their time arguing mining cases, defending clients, or dabbling in politics. The firm of Hudson and Slaymaker in Silverton prepared mortgages, collected house rents, drew up judgments and papers on mining sales, helped settle accounts, secured mine patents, and answered inquiries on such topics as land and mining values. For this they charged various fees. One disgruntled and "not very flush" client complained that he was willing to pay liberally for what they had done, "but it surely looks quite exorbitant to charge me $25 for simply dismissing that case of mine."

The towns, even some of the camps, attracted breweries, which dispensed a favorite beverage of both the miner and nonminer. Few were as fancy as the Silverton Brewery. It promised the "finest beer ever made" in any Rocky Mountain brewery and served ale in a brewery garden that featured entertainment and a lunch counter stocked with all the seasonal delicacies. The management pledged that law officers would be in attendance to prevent any disturbances.

Budding maturity accelerated demands to control lawlessness and rowdiness, much of which originated from too frequent attendance at the friendly saloon. One notable exception occurred when Butch Cassidy, Matt Warner, and a couple of friends daringly robbed Telluride's San Miguel Bank of $20,740 on Monday morning, June 24, 1889, to launch their criminal careers. State, county, and local law enforcement agencies almost totally handled law breaking; rarely was justice taken into private hands under the guise of vigilantism. Eben Olcott noted in 1880 that Silverton contained "such a lot of bad characters" that townspeople formed a vigilance committee to rid themselves of the plague. Over in Ouray in 1884 a particularly brutal murder of a young girl roused the citizenry to hang the guilty parties. The *Solid Muldoon*, while calling it a sorrowful event, did not condemn the committee, which it described as made up of the most conservative, upright citizens. Crimes of physical violence, never numerous, declined, and the most common were those typical of an urban situation— house breaking, burglary, and such misdemeanors as "fast driving." This last seemed all too frequent, deplored Dave Day from his *Solid Muldoon* office, after watching a group of children and a lady with a baby carriage barely escape injury. Main Street was becoming safer and the residents were proud of it. "Our people are as law abiding as any in the world," proclaimed the *Red Mountain Review*, April 14, 1883; thugs and roughs "will do well to keep away from our camp."

Even more heartening for the consumer was a broadening selection of products and a decline in the cost of living, nearly matching the prices of Denver and other Colorado mining camps. Flour in Ouray dropped from $12 per 100 pounds in 1877 to $4.50, and giant powder from

A stage ride on the now widened Ouray-to-Silverton road was a thrill. Miners traveled to Ouray for a dip in the hot springs, which were thought to cure rheumatism, indigestion, alcoholism, and similar ills. Actually, the relaxation and a chance to get some home cooking were probably just as beneficial.

$1 to 35¢ per pound. A dollar purchased three pounds of steak, fifteen pounds of potatoes, and two pounds of onions, and, in Silverton in 1883, left some change for candy besides. For a Thanksgiving dinner $2 provided either a turkey or two cans of oysters and left the purchaser 25¢ to buy some trimmings. A young man could treat his best girl to a sirloin steak dinner at the Board of Trade Chop House in Rico in 1882 for 50¢. The reader must remember that wages of $2 to $4 per day were standard.

Rather new to Main Street and hustling to make up for lost time was the insurance agent. Many companies would not insure miners, considering their job too risky, and fraternal organizations sponsored policies for their members. Thus James Robin, agent for the California Insurance Company, spent most of his time insuring buildings. The policy for a one-story frame house valued at $1,200 cost $15. For the merchant $1,000 worth of stock could be insured for $15, provided that he agreed to the company's requirement that he not keep on hand more than five barrels of kerosene or 25 pounds of gun powder, to be handled and sold in daylight only. This seemed to be a wise precaution and the rates do not appear excessive in those high-risk fire zones of the core community.

Somewhere on the crowded main street of any settlement that pretended to renown sat the newspaper office. Most towns aspired to the status it signified, and newspapers thrived in the eighties. Ironton had one, Red Mountain five, Ouray four, Lake City five, Rico six, and Ames one, to mention a few. Some failed to last a season, others survived a few years, and a few the whole decade. They provided the same services as before and were just as enthusiastically received. The first copy of the *Animas Forks Pioneer* in June 1882 sold for $25, the proceeds going to the school fund. The *Pioneer* was typical of the papers of its day. Each issue carried local and mining news, ads, short stories, and mineral applications, which were required to be published to fulfill government regulations and thus provided an economic windfall for small camp newspapers. Struggling for four years in the "mining centre of the San Juan country" and at a higher altitude than any "newspaper in the world," the *Pioneer* editor and press moved to Silverton in 1886. Precipitating the move were lack of local development, small or nonexistent profits, and the enticements of a more prosperous town.

In its first year the *Pioneer* went through four editors, one of whom was the indefatigable Gideon Propper. His career illustrates the restlessness, or the opportunism, of the San Juan newspaperman. Gid Propper started working in the shop of the *La Plata Miner* in 1879-80. He "moonlighted" by writing San Juan articles for the *Mining Record*, then became editor of the *Pioneer*. Following a tour with Telluride's *Mining News*, a "spicy" paper, a admitted rival, Gid went on to the *Red Mountain Journal*. Sometime in between he found a moment to edit the Lake City *Silver World* and work on two Silverton papers. A competitor wondered aloud, "How do you hook on Gid?" and gave him a tip of the hat for being able to "rustle more mining news" than anybody else. Whether Propper was a victim of an incurable case of wanderlust or was simply unable to get along with the owners has been lost to history. So, unfortunately, have been most of his writings. Propper had a way with words; a fellow editor commented one time that Gil thought "Colorado without whisky would be like hell without brimstone." So much for prohibition!

Newspaper editorships offered an avenue of adventure and success for enterprising individuals. The press was also a force in establishing or breaking down local values via communication. No San Juan paper or editor better exemplified these statements than Dave Day, the region's best known editor during the 1880s, and his *Solid Muldoon*. Feisty, combative, outspoken, talented, fearless, quotable Dave Day jumped into his element in Ouray. He took on everybody and everything; as a Democrat in Republican territory, he found much to debate. His paper and its caustic comments attained fame throughout Colorado, and he gained an enviable newspaper following. Young Mary Mott, growing up in Lake City, found that the *Solid Muldoon* and *Police Gazette* were taboo in her home; when she occasionally found a "racey" copy, she avidly read it. Others did not have to go to so much trouble, nor should the two papers have been bracketed together, because Day put out a family paper, including such ditties as

> Several plates of ice-cream
> And a piece of cake
> Make the finest kind of
> Modern stomach ache.[6]

Yet another feature common to Main Street was a hotel, but not one like Eben Olcott found in Silverton: "Am in a hotel where they open their eyes at your wanting any better place to wash than the common sink of their office. They think you are stuck up if you ask for a bed alone not to say a room." Fortunately, this type was gradually phased out in the mining towns, to be replaced by ones such as the editor of the *Dolores News*, February 18, 1882, recommended: "Good hotels are a very handsome ornament in mining communities, the first thing a newcomer looks for." None was better than the pride of Ouray, the Beaumont Hotel, which opened in July 1887, with a "grand ball and food from the 'finest dining room in the state without any exception.'"

More often now the stage or carriage that rolled up to the Beaumont's door deposited tourists. Tourism was emerging as a selling point for the San Juans, with Ouray their brightest jewel. As soon as it reached the town, the Denver and Rio Grande extravagantly promoted the streams of pure cold water, the hot springs, the fishing and game, the mines, and the beautiful aspen groves. "The moral of all this is, therefore, for health, wealth and the grandest scenery on earth, visit OURAY." Other communities refused to let Ouray grab all the attention. Five of them advertised in the *Tourist's Hand Book* which appeared in 1885. A traveler would "never regret" a visit to scenic Silverton, and Wagon Wheel Gap promoted itself as the favorite sporting ground of lovers of the rod and gun.

Wagon Wheel Gap was now confronted by several rivals to its status as a tourist and health resort. One was obviously Ouray, with its scenery and hot springs; another was Trimble Springs, situated on the D&RG line in the valley north of Durango. Trimble Springs' waters were reported to be the best in the world for rheumatic troubles and good for "chronic difficulties of all kinds." This resort soon emerged as a favorite vacation spot for Durangoans and miners from throughout the area.[7]

The miner who visited Wagon Wheel Gap or Trimble Springs longed for relief from his aches and pains; relaxation may have been his primary goal, however. In the eighties a larger selection of outlets for recreation was available to him, many of the possibilities more refined than previously. Theaters or opera houses graced several of the towns; others converted any available room, from halls to skating rinks, into passable quarters for a traveling company. Roller skating caught the moment's fancy and rinks, which occasionally sponsored races among local champions or town teams, became common. Betting on the outcome whetted interest, as it did in foot races, boxing matches, dog fights, horse races, and baseball games. Most communities fielded a town nine, and claims to being the champions of the San Juans came quickly. "Our Boys," proudly announced Telluride's *Evening News*, July 21, 1884, "are the champion baseballists." The words might have been quaint, but the attitudes have not changed. The gentleman who umpired, observed a San Juaner, "appeared to be a very nice fellow but some of his decisions certainly were very rank." Improved travel conditions and enlarged audience potential lured an occasional circus into the region. Dave Day wrote in 1889 that nothing would offend the most fastidious. He thought the pair of trained burros the best act but also liked the performing horses, trapeze, and contortion acts. It was a bit early and the San Juans too isolated for lions, tigers, and elephants.

Organizations evolved to fill the needs of everyone, and San Juaners joined them eagerly. Social, bachelor, and literary clubs, dancing schools, guard companies (local and Colorado), band and orchestra—name it and they had it. Loosely organized and spontaneous, some of

School—the children might have abhorred the thought of it, but a building such as this at Ouray was looked upon as a community achievement and the mark of a modern town. This 1885 dedication turned out the fire laddies, GAR, and other lodges for a parade.

them lasted only briefly. Others proved more permanent, like the Durango Social Club, with dues of $25 per year, club rooms, by-laws, and a vote on all initiates (three black balls meant rejection). Fraternal organizations thrived; the Grand Army of the Republic corralled the veterans and the Odd Fellows and Masons were only slightly less popular. Christmas-New Year and July Fourth remained the big holidays, their celebrations acquiring more polish and grandeur as time went on. Red Mountain's Fourth in 1887 consisted of a grand parade, orations, races, tug of war, fireworks, and an evening ball. People from the whole area attended.

Unlimited diversions awaited both the "outdoor" and "indoor" person. In one bitingly cold week in January 1885, Silverton offered checkers, cards, parties, sleighing, dances, and saloons. Skiing, on primitive skis usually about ten feet long, won growing numbers of adherents. It was, confessed an addicted practitioner, a most enjoyable and beneficial sport once you "successfully" learned to control the skis and became accustomed to jumps and turns. In this day before manicured slopes, trees loomed as the biggest obstacle.

During the winter, indoor activities attracted the less hardy souls. Saloons, dance halls, and other elements of the red-light district tempted the free and footloose. The prolific saloons quenched thirsts in abundance. "Ladies of the evening" promenaded their wares almost as freely, though their appearance was often seasonal in the smaller camps. They came with

— 86 —

summer and disappeared before the winter snows. Census takers found only a few of them willing to admit their occupation; personal sensitivities or oversight caused many to be overlooked. The Matties, Belles, and Annies were predominantly American born, young, and single, with a tantalizing percentage (33 percent) either married, widowed, or divorced. As a group, the "daughters of prosperity" (an apt euphemism) had the highest divorce rate (8 percent), a dry statistic that failed to determine if this was a cause or effect of their occupation. The "Scarlet Daughters" evoked more sermonizing and Victorian sentimentality than any other single element in the community. "Another gone wrong in life, over whose acts we drew a mantle of charity, hoping before him who rules she may find the forgiveness for her fall on earth for which she prayed with her dying breath," eulogized the *Morning Durango Herald*, March 29, 1887. Suicides were prevalent in the ranks and usually served as inspiration for a journalistic moral lesson. When Grace Marsh took a lethal dose of morphine in Silverton, however, the coroner's jury, on examination, discovered she was still alive. They promptly adjourned until afternoon, when they came together again and found she "was really deader."

Typically, prostitutes found themselves becoming steadily more segregated, if not by city father fiat, then by social pressure. Local Silvertonians were shocked when a "disreputable woman" rented a house in a reputable neighborhood. That such a proceeding was tolerated by city authorities was a disgrace to the town's fair name, complained a neighbor. Mothers also worried about the possible effects on their children of seeing the "chippies." Ouray tried to resolve this by forbidding them on the streets in daylight.

Against the red-light district the church stood firm, though its community-wide role declined as new activities and clubs superseded it. Denominationalism persisted, and it was always difficult to involve a large share of the men. Clarence Mayo explained to his sister that, while there were better cornetists, he was asked to play a solo because "I'm the only one who keeps in respectable society." Ministers and their role barely changed from the seventies. It was hard to go to church, one visitor wrote his wife, because the pastor was so disappointing. The Rev. James Gibbons, Catholic priest serving Ouray and Silverton and numerous other stations, commented that it took good will and courage to succeed; on his first trip to Telluride he slipped on ice, almost sliding over a cliff. His experiences resembled Darley's, yet in spite of these troubles, the church fought hard for the spiritual and temporal reform of the community. "Down with the devil," shouted the *Lake City Mining Register*, while cheering on the combined efforts of local churches and Good Templars, who were holding a temperance revival.

A resident of Sherman grumbled that his community had neither church nor school, a common concern. School buildings were erected more frequently than churches in the smaller camps. Many looked like Capitol City's one-room frame building, which its proud partisans felt was substantial, warm, and well-lighted. The same problems continued to plague the schools—lack of finances, qualified teachers, and community support. Grammar schools appeared sufficient to taxpayers, who feared the rising educational costs of going beyond that level.[8]

The mounting cost of municipal government also aroused fears of increased taxes. By the mid-1880s all the mining towns and a few of the larger camps acquired "city fathers"; the rest got along as best they could, eschewing such trappings and expense.

Ouray, Lake City, and Silverton city governments completed their first decade during the eighties. Having passed through the growing pains of youth, their meetings, except in a crisis, were marked by more deliberation, less urgency, and more predictable responses. Overall, the services they provided were more routine in nature. In a nutshell, small town urbanization replaced youthful acceleration. A brief survey of Ouray's efforts serves as an example.

In the period from January 1883, through February 1884, meetings were held at least monthly and only one had to be canceled for lack of a quorum. Except when there was a smallpox scare in May 1883, special meetings were infrequent. Taxes, water mains, construction of a bridge, rain damage, the town abstract, and a new fire hose caused the council some anxious moments. A regular meeting, such as that of August 10, 1889, included a variety of routine matters, like monthly reports, salaries and expenditures, approval of license bonds, and petitions (three this meeting—purchasing some city land, addition to a business building, and opening a new street). More interesting were instructions to the town attorney to draft an ordinance prohibiting livestock running at large. Ouray's new-found respectability would not permit this vestige of a more primitive time. Other signs of change were evident in ordinances regulating gambling, disorderly houses, shows, exhibitions, and amusements, and one controlling vagrants and beggars. Petitions from worried citizens originally prompted the city fathers to inspect gambling and other social evils, which meant dance halls. The definition was expanded to include all shows and amusements in the ordinance. Not all petitions proved successful; one for Sunday closings of business was tabled.

Another sure sign of municipal growth was construction of a water works and the laying of mains along the principal streets. A water works department emerged, and license fees for plumbers and water rates were set. Before the decade was over, additional municipal water bonds had to be approved to pay for a larger reservoir and more water mains. Repairs, citizens' complaints, bond payments, and delinquent water bills bedeviled the council on numerous occasions. So did complaints about rubbish, which resulted in attempts to clean up Ouray. On a more agreeable note, the contract between the city and the Denver and Rio Grande for a right of way through town was approved; few complained about this.

Concomitantly with increased services and responsibilities came a rise in governmental expenses. Expenses for the year ending March 31, 1883 amounted to $8,864; they exceeded $20,000 for the year starting April 1889, excluding interest on the water bonds. Revenue came primarily from business licenses (liquor leading the way), water licenses, county payments, and a property tax. While it was a better revenue-raising system than Ouray possessed earlier, it still failed to generate as much money as needed, particularly to meet the water bond payments. Nor was complete enforcement of the statutes that appeared on the books secured. Although prostitution was banned in town and within three miles of its boundaries, nevertheless it flourished. Perhaps the fines levied against it provided a premeditated but clandestine source of revenue; the available sources discreetly avoid such an admission. Social needs and public morality clashed on this question; the former continued to win out. Law enforcement never pleased everybody; the marshal, the hardest worker of the town officials, could not reasonably be expected to do the variety of jobs assigned him to everyone's satisfaction. As the most obvious contact point between citizen and government, the marshal attracted the most attention and comment.[9]

Telluride and Rico, meanwhile, endured the throes of government organization, then stabilized, and by 1889 appeared much like their elder neighbors. There were more special meetings, ordinance discussions, meetings canceled for lack of a quorum, and government procedural issues than occurred in Ouray. Still, Telluride had time to worry about trash, pass an antipollution ordinance related to the watersheds, and discuss promoting the community in a state history. Rico created a water works, became embroiled in a fight with Dolores County over taxes, formulated a budget of $8,000, and organized a fire department. A Ricoite promptly sued the town when his building was destroyed on order of the fire chief to help stop a fire. An ordinance was even passed charging those who sold oleomargarine a special $25 license fee.[10]

Local reaction to city government was seen in letters to the papers, editorials, and in heated discusssions during board meetings. Regrettably, the minutes remained uniformly laconic about what went on, mostly cryptic notes and mention of petitions. Expenses vexed everybody. Rico voters thought the mayor and trustees ought not to be paid, but they got nowhere, while Lake Citians were more successful in paring town salaries during the hard

Every mining town (and even the camps) wanted a newspaper. Silverton had the San Juan Herald *and a host of others during its mining prosperity. Each one promoted and defended local interests and attacked anyone who dared question that its mining district was the greatest.*

times of the mid-1880s. Silverton's city fathers nearly found themselves in court for failing to enforce ordinances, and Animas Fork residents grumbled about but went along with the town ordinance mandating brick chimneys. Before the conclusion is drawn that all reactions involved grumbling, let it be known that Telluride's board and mayor were congratulated in 1882 for handling the town affairs in such a satisfactory and businesslike manner. They had done all in their power to "promote the interest and welfare" of the town.[11]

Politically, it was becoming harder to type the San Juaners, mavericks that they were, who voted for the man as often as the party. The various governors' races are good examples. Much to Dave Day's delight, his party captured the county three out of five times, yet only twice did all the San Juan counties give one candidate a solid majority. In 1882, Democrat James Grant, and in 1888 Republican Job Cooper, swept to victory, the candidates' personalities affecting the outcome more than the two parties' platforms, which varied only marginally. The San Juans backed the Republican presidential nominees during the decade, however. In local races for town offices or the state legislature, very few candidates survived two terms, whether by personal choice or voter preference.

These, then, were the San Juaners and their communities in the 1880s. They cleared, carved, and urbanized for twenty years. Much of the building was completed; refinements and possible decline awaited. Back on September 10, 1881, the editor of the *La Plata Miner* asked that every effort be made to make Silverton a desirable place for residence. He specified that this should include good schools, churches, civic societies, and water and gas works. The same challenge echoed throughout the San Juans. In the eighties time and attention had turned and the frontier image faded.

[1] Ingersoll, "Silver San Juan," page 697. Olcott to sister, November 4, 1880, Olcott Collection.

[2] The following description was based on the original census returns of the federal census of 1880 and the Colorado census of 1885.

[3] See the *Dolores News*, May 13-27, 1882; *Solid Muldoon*, November 16, 1883 and March 11, 1887; *Red Mountain Pilot*, March 24, 1883; *Animas Forks Pioneer*, June 24, 1882, July 24, and August 7, 1886. *Silver World*, February 25, 1882. Carolyn and Clarence Wright, *Tiny Hinsdale of the Silvery San Juan* (Denver: Big Mountain Press, 1964), pages 144 and 180.

[4] William Pabor, *Colorado as an Agricultural State* (New York: Orange Judd Company, 1883), pages 142-144. Peter Archdekin Interview, Bancroft Dictation, Western History Collections, University of Colorado. *Telluride Republican*, July 3, 1886. *Solid Muldoon*, June 17, 1881. Ingersoll, "Silver San Juan," page 691. Eben Olcott to sister, May 22 and June 10, 1881, Olcott Collection.

[5] Dave Wood correspondence, David Wood Collection, Colorado State Historical Society. *Solid Muldoon*, January 8, 1886. *The San Juan*, June 2, 1887. *Animas Forks Pioneer*, February 27, 1885 and July 10, 1886. Mears's account, Journal No. 1, Bank of San Juan (Durango). For Dave Wood see Frances E. Wood, *I Hauled These Mountains in Here* (Caldwell: Caxton, 1977).

[6] *Solid Muldoon*, November 5, 1880. The preceding section on Main Street and its inhabitants was taken primarily from local newspapers, supplemented by the following: Signature Book of the First National Bank of Durango; Hudson and Slaymaker Collection, California Insurance Company policy book, and F.C. Sherwin Customer Purchase book, San Juan Historical Society; Eben Olcott Collection: Mary K. Mott, "At Lake City and Telluride," *Pioneers of the San Juan*, volume 2, page 95. Matt Warner, *The Last of the Bandit Riders* (New York: Bonanza, 2nd Reprint), chapters 15 and 16.

[7] *Durango Record*, September 10, 1881. *The San Juan*, June 23 and July 28, 1887. Olcott to wife, September 9, 1880, Olcott Collection. *Denver and Rio Grande Official Local Time Tables* (Denver: Denver and Rio Grande, 1888), pages 103-105. *Tourist's Hand Book to Colorado, New Mexico and Utah* (Denver: 1885), pages 41-50. For the saloon in general see Elliott West, *The Saloon in the Rocky Mountain Mining Frontier* (Lincoln: University of Nebraska, 1979).

[8] *Lake City Mining Register*, November 23-December 7, 1883. *Silver World*, June 19, 1880. James Gibbons, *In the San Juan* (Chicago: Calumet Book and Engraving Company, 1898), pages 7-23. Clarence Mayo to sister, April 9, 1882, Clarence Mayo letters, Henry E. Huntington Library. Eben Olcott to wife, May 30, June 6, 1880 and February 6, 1881, Olcott Collection. *Resources and Mineral Wealth of San Miguel County, Colorado* (Denver: Publishers' Press, 1894), pages 50-52.

[9] Ouray Books Five and Six, Minutes of the Proceedings of the Board of Trustees, January 9, 1882-December 31, 1889.

[10] Columbia (Telluride) Books 1 and 2, Minutes of the Board of Trustees, June 5, 1880-September 21, 1886; Telluride City Record, Book 1, April 16, 1888-December 31, 1889. Rico Book of Ordinances, 1880-89; Book 1, Minutes Board of Trustees, December 25, 1886-December 31, 1889.

[11] *San Miguel Journal*, February 4, 1882. See also *Animas Forks Pioneer*, July-August 1882. *Dolores News*, February-March 1882. *La Plata Miner*, January 31, 1885. *Silver World*, March 1883.

T he 1890s were for the San
Juans years of transition and
years of potential finally
realized. Two decades had passed
since permanent settlement and
mining had come—a long time for
a mining region to wait, a long
time for a mining region to produce. The wait now proved
worthwhile; the bonanza forecast
in the seventies, tantalizingly
hinted at in the eighties, became
reality in the nineties.

After years of deficit mining, the
San Juans turned a profit. The
creation of a costly transportation
system was completed, as was the
establishment of an urban core of

Golden Backbone of the Silver Mountains

camps and the initial development of the mining districts. Mortality of the smaller mines rose,
victims of unprofitability, leaving behind revenue-producing or more promising properties. By
the end of the nineties the major mines were open and operating, with the main expenses
being ones of exploration and development. Although statistics do not make exciting reading,
they can be enlightening. For example, the San Juans in 1890 produced $1,120,000 in gold
and $5,176,000 in silver. Nine years later the figures were $4,325,000 and $5,377,000. A quick
perusal shows gold mining making a strong gain, as did total production. Silver held steady
even with the crash of 1893, demonetization of silver, and a lower price. Production in ounces actually rose to counterbalance price decline. However, it should be pointed out that by
1899 silver was more a by-product of gold mining than strictly its own master. The San
Juans, long a two-metal region, were in a far better position to ride out the mining traumas of
the nineties than an exclusively silver district.

A breakdown of production figures by county makes the growing dominance and signficance
of the big three towns—Telluride, Ouray, and Silverton—even more apparent than in the
1880s. Eighty-seven percent of the total in 1890 came from these three, and in 1900, 71 percent; add to that percentage Creede's total and it runs up to 95 percent. The rest of the San
Juans were dropping by the wayside; mining was passing them by and investors looked
instead to where the profits flowed.

Creede—"It's day all day in the day time, and there is no night in Creede," sang its
poet-newspaperman Cy Warman. Creede, overpromoted and exploited, was the only San
Juan mining district that ever received the same acclaim and publicity as Leadville and
Aspen. Some other camp or district always seemed to grab the headlines away from the likes
of Ouray or Lake City, but not Creede. Creede came roaring into the spotlight in 1891-1892,
the "last" of the Colorado silver camps. Unlike earlier San Juan excitements, Creede was
easily accessible; the Denver and Rio Grande nearly touched its back door when the rush
started. Journalists, newspapermen, and sightseers took full advantage of the opportunity "to
see the elephant," as they once said of the California gold rush. Fortunately, a mineral basis
existed for the stampede, and for a while Creede glistened like a new Leadville. Even another
county, Mineral, was carved out to satisfy its wishes.

Prospectors wandered in and around the Creede region for years; it was just off the old road
from Del Norte, via Stoney Pass, into Silverton. One veteran prospector in particular, Nicholas Creede, gained faith in the district, and his discovery of the "rich from the grass roots"
Holy Moses mine in 1889 persuaded others to come. Along with those others came inves-

tors, like David Moffat from Denver and Thomas Bowen, who had already made a fortune at Summitville. They bought the Amethyst and Commodore mines, and Creede, unlike other San Juan districts, secured excellent transportation, an abundance of investors, and had producing mines within two years. No wonder, then, that Creede caught the public's attention.

Creede's star twinkled brightly over the San Juans, briefly and audaciously challenging Cripple Creek, Colorado's other new mining bonanza. The best years proved to be 1892-1893; after that, except for a revival in 1897-1900, it was downhill for this second Leadville. The declining price of silver hurt, as did an early encountered water problem. Both of these obstacles were mastered in other San Juan districts. What really hurt Creede was the rapid decrease in ore value once the first rich pockets played out. Combined with the silver question, it was enough to foredoom Creede to a frenzied, short boom. Commenting on Creede, a Howardsville resident, J.S. Robinson, prophesied in 1892 that the mines would support a small town and the boom "will quickly run its course."[1] Other San Juaners were less sage and more envious.

Creede was the most noted San Juan district, though not the only one, to be dealt a body blow by the issue which Coloradans and western silver interests felt to be the fundamental one facing the United States: the plunging silver price and what the government planned to do about it. Not a new issue, it dated back to the 1870s, when the government stopped coining silver. Though unrelated, this occurred at the same time as a collapse in the world silver market. Together, these events stunned western silver miners and produced rising demands for help. The result was two bills, the Bland-Allison Act, 1878, and the Sherman Silver Purchase Act, 1890, both of which provided for government silver purchases. Westerners, meanwhile, sought something more than just a simple purchase remedy; they wanted a guaranteed price. San Juaners, emotionally swept up by the issue in the 1880s, had the urgency of the matter driven home by the financial crash of 1893. The crash, then depression, the worst the United States had yet suffered, brought hard times that were too easily and erroneously blamed by many easterners on the Sherman Act, which they claimed depleted the treasury of its gold to buy silver. President Grover Cleveland agreed, and together they had enough votes to repeal the act during a special session of Congress in 1893. The price of silver promptly plunged to 60¢ per ounce, a drop of 20¢ in four days. Colorado, hard hit, fought to redeem silver, and the silver question boiled. What silverites wanted was a higher price, based on parity with gold at 16 to 1, and the assurance of a government market, if the world market could not absorb production. Economical and political ramifications appeared immediately, only the former being of interest at the moment.

Mines shut down or curtailed production, laying off miners and adversely affecting Colorado's entire economy. For those fortunate enough to keep their jobs, wages were reduced, with grumbling the only safe reaction; a strike would give jobs to the unemployed, who would willingly seize at the chance. Business throughout the San Juans declined sharply, whether in the neighborhood store or on the railroads, followed by inevitable failures and bankruptcies. Gradually at first and then more rapidly men moved on, but where were they to go, except perhaps Cripple Creek? The gloom grew darker during the winter of 1893-1894. Otto Mears spoke for many when he wrote that it was impossible to work silver mines with present prices, and the shutting down of mines meant the closing of many enterprises affiliated with them.

Through it all a light could be seen, not a silver lining, but a golden one. The collapse of silver sparked a new search for gold, a search that in the San Juans softened hard times and eventually stabilized the regional economy. Many of the larger mines, particularly Telluride's, which yielded both metals, had been less drastically hurt. As John Porter observed, the decline was bad, but it was, after all, only a few dollars per ton. Another San Juaner wrote the *Financial and Mining Record*, April 30, 1892, a year before the real holocaust hit, that "there is no great evil without some good"; gold mining and prospecting were being stimulated. While this was true for the mines with gold, it proved deadly for the silver-only producer. By mid-decade the changeover was noticeable. Gold mining was coming on strongly, while silver production came to be centered in bimetal districts, where the emphasis simply changed from silver to gold. The result pushed San Miguel, Ouray, and San Juan counties, in that order, into the forefront, where they remained for two decades. The other result, less quickly perceived though equally important, was the dominance of large companies in San Juan mining. Ouray mining reporter and old-time San Juaner Chauncey L. Hall saw clearly what was happening when he commented that nearly all the promising mines were going into the possession of "heavy" capitalists, leaving the man attempting to mine in this country without capital, "as helpless as a boat in midocean without rudder or compass." Hall concluded:

> Money is the lever of power that moves the business world and money is the only power that will turn these mountains inside out and collect their inestimable riches.[2]

What did this mean in terms of San Juan mining? Several examples will illustrate why the small miner found himself frozen out. The Silver Lake mine, gold and silver, between Howardsville and Silverton, developed steadily in the nineties under the careful guidance of owner Edward Stoiber. By 1899 the physical plant included the mine itself, a boarding house, mill, employees' homes, office building, power plant, nearly three miles of tramway, and a telephone line connecting all main offices. Stoiber's property of 1,014 acres included 116 claims, and he employed 281 men, 160 of them miners. In the summer of 1899, the mine produced 6,000 tons of crude ore per month, which was reduced and concentrated at the mill, then shipped to the Durango smelter.

A rare find in mining history is assay records, such as those for Silver Lake, which have been preserved from the 1890s and offer a revealing look at operations during this crucial decade. The ore quality held steady, even with the silver price problems, and the mine showed strong silver content and moderate gold (usually less than one ounce of gold and ten of silver to the ton). Only rarely were rich pockets found; one was the 148-ounce silver ore mined in 1895. Lead offered no salvation and was recovered only haphazardly in concentrates. Tailings, unprofitable to work but sometimes carrying an ounce or more of silver, were dumped unceremoniously into the lake.

Electric power was used throughout, and underground "air drills" accelerated mining. The tramway was one of the wonders of the region and one of the longest on the American continent. It stretched from the mine terminal to the mill, directly beside a railroad spur. Stoiber wisely reinforced the tram towers with V-shaped bulkheads and masonry piers to protect against giant snowslides, which thundered by. The tramway served many purposes, hauling up all the boarding house stores, mine supplies, hay, timber, rails, and coal, an arrangement, Stoiber carefully pointed out, that provided economical handling of such material. Thoughtful of his men, he tried to provide the best possible accommodations and food in order to hire and maintain "a superior class of miners and mill men," in spite of a disagreeable winter climate and high altitude.

Across the mountains above Telluride sat an equally impressive and more famous mine, the Tomboy. Fortunately, this property was examined in 1896 by James D. Hague, who twenty years before curbed John Moss's dreams in La Plata Canyon. His records have been preserved and provide an insight into San Juan mining. The Tomboy had been worked off and on for years—then, like Silver Lake, it emerged as a major producer in the 1890s. Hague visited the Tomboy for the Exploration Company of London, which subsequently purchased it and appointed him president of the Tomboy Gold Mines Company. Hague's report showed that the mine produced $1,250,000 between January 1895, and November 1896. He explained that no production figures could be precisely verified before that date. Mining, mil-

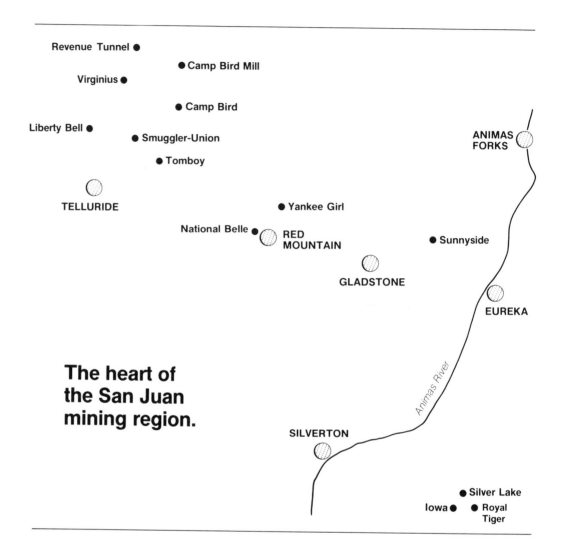

Revenue Tunnel ●

● Camp Bird Mill

Virginius ●

● Camp Bird

Liberty Bell ●

● Smuggler-Union

● Tomboy

TELLURIDE

ANIMAS
FORKS

● Yankee Girl

National Belle ●

RED
MOUNTAIN

● Sunnyside

GLADSTONE

EUREKA

**The heart of
the San Juan
mining region.**

Animas River

SILVERTON

● Silver Lake

Iowa ● ● Royal
Tiger

ling improvement, and general expenses equaled $648,105, leaving $601,895 profit. Annual taxes amounted to nearly $6,000, a figure well beyond the means of many San Juan mine owners, as was Hague's $5,000 fee, plus expenses.

Impressed with what he saw after a thorough inspection, Hague recommended the purchase. Hague took time from his work to describe to his wife some of the problems of high elevation mining. Caught in snow and cold, he confessed, "this is good enough for those who like winter and 13,000 ft [actually about 11,500 at the mine]." The elevation proved taxing; he tired very quickly and the slightest effort made him "puff and blow." Up before 7 a.m., he was soon thereafter into the mine, where climbing 300 to 400 feet on steep ladders reminded the 60-year-old mining engineer that he had lost a step or two since his La Plata Canyon days. Crawling through small holes, usually "very dirty and nasty," he commented to his wife, "you can imagine what a nice time I am having." Regardless of the discomforts, Hague retained an eye for beauty; the beautiful moonlit nights and "bright dazzling sunshine" on the fresh snow caught his attention.

Under his direction (he usually visited the property once a year), the Tomboy produced well, even spectacularly. He worked especially hard to find a competent "mining captain" (underground superintendent) and general manager; in Hague's absence, they ran the operation. During the next two years, adjoining claims were purchased, a new bunk house built, Ptarmigan Lake acquired to furnish water for the mill, new equipment installed, and 300 men employed. By the end of 1899, Hague had left the firm, but he had helped turn the Tomboy into "one of the world's great gold mines," one of the "big industries of Southern Colorado."

Just around the corner and a mile or so northwest from the Tomboy sat the Smuggler-Union, located high in Marshall Basin, a mainstay of Telluride mining in the 1880s. A brief glance at its history in the 1890s shows mining's tribulations. Two adjoining claims were consolidated in 1891 to form the mine, an obvious move to save expenses and lower costs. John Porter, who believed consolidation to be a wise move, emerged as president of the company, which included several other claims besides the two mentioned. In 1892, the company was staggered when ore values in the Smuggler dropped off alarmingly. Porter promptly advised stockholders "not to be alarmed at the fact of no dividends"; their reactions are unknown. Part of the reason for subsequently passing the dividend could be traced to expenses, as Porter put the property in shape for large-scale mining by building a tramway and carrying on development work to open new ore sources.

The declining price of silver also cut into profits. The year 1893 proved just as bad for the beleaguered Porter, who suffered great anxiety over heavy expenses but boldly stated "it would be a sin" to stop. The stockholders stuck by him, and in late 1894 all were encouraged by the finding of rich gold and silver deposits.

Only further expenses in building boarding and bunk houses and tram terminals characterized 1895. Profits disappeared in the expansion. The Smuggler leased the neighboring Sheridan, whose owners were not too sure that their property was being worked to the best advantage, another headache for Porter as he tried to pacify them. Finally, in 1897-1898, the corner was turned and the Smuggler-Union operation rivaled the Tomboy's. At this point, in April 1898, the New England Exploration Company purchased the property, continuing operations on a large scale.[3]

These three mines show the problems, pitfalls, expenses, and methods needed to carry on profitable mining in the San Juans. Mining had come a long way in just the past few years. Absentee, corporation control came to dominate in two of the three, an indication of what was happening elsewhere. The profits might be great; expenses assuredly would be. Trained personnel now managed the properties for owners who might never see the inside of a mine, nor even the San Juans, except in photographs.

Obviously, all mining was not conducted on this scale; most mine reserves could not have justified the expense. Experiences of the numerous, less well-financed owners proved more like that of Malcolm Downer, who in 1896 opened a mine in Yankee Boy Basin, above Ouray, that he called the Bimetallist. High-grade gold and silver at the surface started Downer off in fine fashion, then values dropped and dampened his enthusiasm. Four years he worked his property, while the silver price declined; mill facilities to work his lower grade ore were generally unavailable and his finances hovered at nearly zero. Finally, one winter, he discontinued his work, "awaiting better-time," which somehow never came.[4]

For the Downers of the San Juans, the mining era was ending. Steadily, unceasingly, the best mines fell into company hands. The individual might find and open a prospect, but his best bet would be to sell eventually. Mining in the San Juans had become big business, with ample finances the key to success.

If finances provided the key, then investors fitted the key to the lock. They traveled into the San Juans in greater numbers now that the mines offered profitable investment potential, and one could journey in comfort almost to the mine portal. Investors sent mining engineers such as Hague or Clarence King, who worked for the North American Exploration Company, or they came on their own, following the wise advice of Otto Mears to a potential investor, "Go examine the property."

Knowledge about mining was no prerequisite, although it helped; ready cash was the only requirement—the San Juaners provided the rest. As a result, many investors leaped before they looked, to their own grief but to the lawyers' profit, as they represented clients trying to untangle mining snafus. The files of William Searcy and Edgar Buchanan, Silverton lawyers, are crowded with correspondence from aggrieved investors. Matters of concern were annual assessments and taxes (when due and why so high), selling their property, parties to lease their mines, securing patents, and steps to protect their property rights. One has to sympathize with them, as their visions of quick wealth faded away; they fell victim primarily to their own greed and ignorance.

Both Americans and foreigners invested. In 1897 George Roberts, director of the mint, discerned some interesting investment patterns. American capital, he observed, controlled nearly all of Creede and Silverton and two-thirds of Ouray, while Rico was at least half foreign investment and Telluride and Ophir higher than that. Reasons for these preferences cannot be attributed to anything except perhaps personal contacts between early arrivers and later comers. Telluride always proved attractive to foreign investment and came to foreign attention early. The English were active in the Red Mountain district, where they were hailed as wise investors. John Bull, wrote a San Juaner, was "dipping into good things"; in fact, their success had done much to "restore confidence" in American silver mining investments. The expertise of the English proved fallible when they eventually got burned badly at Red Mountain.

Red Mountain began the nineties as one of the major Colorado mining districts. It retained that position for several years, led by the Yankee Girl Mine, which produced $413,000 from July through November 1890, with expenses running considerably less than half that. By 1897 the Yankee Girl and much of Red Mountain had closed down. English investment had run aground on the rocks of a collapsed silver price, much lower grade ore, and high mining costs, added to the trials of winter and acid-impregnated water. By decade's end, leasers picked over the remains and Red Mountain was dead as a major producer. Crusty Dave Day had been leery from the start, editorially wondering why those "Yankees" would allow the "chappies" to gobble those mines.

The rest of the San Juans offered a study in contrasts but little encouragement except to the most optimistic. Plenty of hope persisted. Lake City and Hinsdale County were making a comeback of sorts, this being the best decade in their mining history. Production, however, only twice topped $700,000. The English were active here, purchasing the Ute and Ulay (note the changed spelling), with only slightly better success than on Red Mountain. Carson "woke up" for a while, an additional town site being laid out on the east side of the divide. The same problems that retarded it in the 1880s stymied it in the 1890s. Even in the successful Telluride district, poor Ophir was rated in 1899 as "still promising," its lower grade mines standing in glum contrast to the wealth just a stone's throw away. And those San Miguel placers, where golden dreams once banded, vanished amid boulders, unexpectedly high costs, and piddling returns.

Rico, predicted as "destined for greatness," experienced its last great year in 1893. The rich ore veins had been exploited and mined out, and, as in Red Mountain and Creede, the plunging silver price spelled doom. Just to the south, the La Plata Mountains stirred with gold fever, encouraging one reporter to say, "in spite of the old saying 'there's nothing there' there are good reports." Nearer the truth was the comment that the district had so far been a failure, "large assays, but the quantity is not there," echoing James Hague's prediction of the mid-1870s. An analysis of what else went wrong would include lack of investment, blunders in selecting mill processes, and want of capital to modernize. The Bear Creek district (called Gold Run) between Silverton and Creede also elicited excitement. Depression-born in 1893, it matured and grew old in a few years, never approaching its vaunted potential.[5]

Alas, for a large portion of the San Juans, their days in the mining sun were behind them, a

"It's day all day in the daytime and there is no night in Creede." Creede burst onto the scene in the 1890s and faded nearly as rapidly as it rose. Two ore trains are getting ready to go into the Commodore mine and so, apparently, is a crew. By 1899 Creede was enjoying its last good days, but it is doubtful the Commodore produced one million dollars that year.

The Virginius mine, at 12,100 feet, was one of the San Juan's highest and also contained what was probably the deepest shaft. Mining was a precarious business. As mining engineer Louis Rickets observed, you needed to develop the property before large capital expenditures were made on a plant, and you needed sufficient capital.

fleeting shadow of yesterday's hopes. For a select few the future held unlimited promise, more dazzling than anything thus far. That mining could return profits can be seen in the cases discussed. Even in the Downer example, though no great profit had been turned, he had mined enough ore to scrape out a living for four years, typical of the smaller operations. These small mines produced merely living expenses for the owner and perhaps a few miners. The opening expenses were high, then cost versus production leveled off, and until the inevitable decline in ore value, the mine operated at this pace. Production on this scale rarely rated headlines or even a story; however, all these little mines contributed a share toward keeping the region afloat.

An 1891 summary listed 500 working San Juan mines, a figure that dropped in the 1890s and, had it not been for Creede, would have declined farther before the 1893 crash. Consolidation, corporation growth, and the switch from silver to gold distinguished this decade. Giving impetus to these developments was an accelerated search for new and improved mining methods, with the money at last available to support the endeavors. As a result, the San Juan mining region pioneered in the development and/or acceptance of improved techniques, none more exciting than electricity.

A few far-sighted San Juaners begrudged the waste of water power tumbling over falls and racing swiftly downstream. J.S. Robinson, writing his Leadville friend Henry Moody, mulled this over and asked, "Could not this water power be converted into electricity?" Indeed it could; the guiding genius was Lucien Nunn, an ambitious Telluride lawyer-turned-mining man. As manager of the Gold King, a mine southwest of Telluride perched high in the mountains, he lost money because of the expense of freighting fuel. Nunn evaluated several means of reducing costs, including the use of direct electrical current, already tried at Aspen. He settled, however, on experimentation with alternating current, which could transmit higher voltage more cheaply and easily than direct current. Up to this point, however, it was all theory. Nunn visited the Westinghouse Electric Company in Pittsburgh and, after a series of experiments, began construction of a plant at Ames. In the spring of 1891, the generator was revved up and electric current raced to the mine and mill. It ran thirty days without stopping, proving the success of the idea, although rumor hinted that the operators were afraid to shut down for fear it might not start again. The first alternating current to be transmitted over a long distance and the first working commercial venture—Nunn had attached two "firsts" to his name. He had been a real pioneer; his fellow San Juaners soon came to realize the total significance of his accomplishments. Many locals came to watch the novelty of the often sparking, smoking fireworks that accompanied the machinery's start on Sunday evening, once operations had stabilized on a regular six-day basis.

The stockholders of the Gold King quickly grasped that this innovation reduced expenses and put the property into profit. Edward Stoiber also understood and installed the largest and most complete multiphase plant for mining in the United States, using water power and coal to run his generating station. The Virginius mine had actually preceded Nunn, but it used its own direct current plant to produce power, at a saving each year which more than repaid the construction cost of the plant. The cost of coal to this mine had run about $36,000 a year in freight. But electricity was not cheap either—the Tomboy spent $1,300 per month for power and light. Nor did all mining men accept the new power; some remained wary. With mining depressed, the initial outlay for equipment and machinery could be staggering. In the end electricity won out, the saving on fuel alone coming as a godsend when the silver price skidded. Electric power also required less attendance and fewer repairs than steam or compressed air engines, there was less danger of fire and explosion, and it was more adaptable and efficient. Who could afford to hold out?

San Juan towns and camps eagerly accepted electricity, Durango as early as 1887, with Creede, Ouray, Telluride, Rico, Silverton, and Lake City not far behind, while others discussed it. There was even talk of an electric railroad from Ouray to Ironton, climbing that

To get tram cable, or any kind, into the mountains was a tedious, difficult task. This mule train stretches well back into Telluride, as it prepares to transport a cable. Just loading and unloading the animals took skill and patience.

The Silverton, Gladstone and Northerly Railroad was completed in 1899 from Silverton to Gladstone to serve the Gold King mine and the little camp there. The assay office of the mine in Silverton also served as a ticket office.

impossible grade that stymied railroads. The plan died, a victim of the crash of '93 and the collapse of Red Mountain mines.⁶ Electricity changed the life of the San Juaner and mining in so many ways it would be futile to attempt to catalogue its total impact. A light came to the dark recesses of the mines and brightened the home parlor, changing the pace and flavor of life. Few denied it was all for the better.

Herbert Hoover wrote that California mining men had "rooted skepticism" about "them college educated fellows," a sentiment many a Colorado old-timer echoed. In the 1890s, trained mining engineers came to be more acceptable. No small factor was the emergence of companies, whose stockholders knew little or nothing about mining and were not sensitive to the feelings of the graduates of the school of hard knocks when it came to tallying profits and losses. Hague and the other mining engineers brought trained hands, a better bet than the old-timer of questionable reputation. Thomas Rickard, later a famous mining editor and writer, came to the San Juans as a mining engineer. Working mostly at Red Mountain and Rico, he also did consulting. A man of principle, he resigned as manager of the Yankee Girl mine when the owners only grudgingly paid their workers. It was a common trick to let the payroll fall in arrears, then close the mine, leaving the employees high and dry. Rickard would have nothing to do with that. At Rico he was offered a $1,000 bribe for a favorable statement to boost stock.

> I smiled, and suggested that he throw his paper-weight out the window. 'What for?' Because he might hit a man on the head that would be willing to make the same statement for less money.

Consulting took him everywhere; during a trip on skis into the La Plata Mountains he traveled at night to avoid snowslides. Other inspections required long horseback trips. Reminiscing years later, he concluded that it had been a "healthy and vigorous" life.

Some mining engineers busily promoted and sold mines. John Farish spent several months in 1891 trying to bring together Isaac Ellwood, a DeKalb, Illinois manufacturer of barbed wire, and a Rico mine: "I am well satisfied that this is the best and surest mining proposition that I have seen for five years, and it is what I have been waiting for before talking with you." Farish cautioned that speculators were turning their attention toward Rico: "It will be necessary for prompt action to prevent prices being run up on us." The price was run up, not as Farish had foreseen, but by a woman part owner, who apparently thought so highly of her share that she set "an outrageous figure" on it. Possessing little hope of persuading her to change her mind, he discouragingly wrote Ellwood in August, "It now seems altogether probable that I will not succeed in getting it at such a figure as would prove a good investment."⁷ The transaction never came to pass and Farish turned his attention elsewhere.

Electricity was not the only aid to the work of the mining engineer and miner. The telephone closed the communications gap, which once existed from mine to mill and from them to town. It also improved the health and assured better safety for the men, now that a doctor was only seconds away. Tramways became longer, faster, and technically improved. The obvious benefits demonstrated in the previous decade paved the way for general acceptance. Skillful use of tunnels to tap veins at lower altitudes (the Revenue above Ouray was a classic example) permitted ease of access, combined operations, and lower operating expenses, all of which helped everyone, from miner to stockholder. A modification in the construction of ore chutes—putting the logs butt end out rather than sideways—seems immaterial, yet they wore longer and allowed passage of heavier rock down the chutes. Power drills came into more general use, especially in the larger operations, while compressors became larger, able to operate more drills. Cages, carried by improved steel cables and run by more powerful engines, replaced earlier hoist buckets. Finally, the safety cage, with a safety brake, overhead roof, and mesh on the sides, provided the miner with a degree of protection never before experienced. Even so, five miners were killed in December 1896, when a cage fell 1,100 ft in the Virginius shaft. The safety catches had not functioned, nor had the engineer and foreman taken proper precautions and exercised their duties correctly. In all, by the end of the 1890s the San Juan miner was blasting and moving more ore and rock per day than his counterpart of a decade before, and he was doing it more easily.

The San Juan millman did not lag behind his mining brother when it came to improvements and modifications. The most prominent of these was the concentrating mill, where low-grade ore could be worked and concentrated into a form that could be profitably shipped to the smelters. In this way low-grade ore was saved from the dump, and the mine's life was lengthened. Consequently, concentrating mills appeared everywhere. Edward Stoiber had one of the most modern plants at his Silver Lake operation; others were found at Creede, Ouray, Telluride, Sherman and Ophir, as well as at the mines near Eureka, at Sneffels, Titusville (above Silverton), and near Ironton, to mention a few. Here the ores were crushed, stamped, and worked to remove gangue and to save as high a percentage as possible.

Another trend was the growing consolidation of mine and mill, exemplified in Telluride at the Smuggler-Union and Tomboy operations. These were first concentrating mills, before they became more refined. Pandora, directly east of Telluride, was the favorite mill site, with trams strung from mine to mill. Improvement in processing came when tests showed that a startling percentage of ore was being lost—over 50 percent at the Sheridan mill; the Smuggler-Union lost nearly 20 percent of gold and 40 percent of silver values down the San Miguel River.

As a result, these companies pioneered in the development of new processes, especially cyanide. First tried on the tailings, where previously unrecoverable ore was recovered, it evolved into cyanide mills by the end of the decade. The Liberty Bell upped its savings to 72 percent, from a previous 58 percent, by combining amalgamation, concentration, and cyanidation. Cyanidation held out great promise, not only to increase saving of ore mined, but to recover values earlier sacrificed.

Smelters continued to be important, despite these new trends, with Durango retaining its lead as the smelter center, though not without rivals. Ouray, which had nearly pined away waiting for a smelter, finally achieved its goal. Hailing itself as a new center, Ouray ate crow when the mill closed, a victim of inadequate ore supply and high railroad rates. Rico fell the same way, a smelter there unable to compete with Durango and Denver. How many failures would have to occur before the evidence became clear? Silverton also came forth to challenge and for a while proved a worthy rival. A smelter that opened in 1894 challenged Durango by cornering Red Mountain's production; then Red Mountain collapsed and Silverton's smelter was dealt a crippling blow, one from which it never recovered.

Durango's San Juan Smelter was the region's largest, and it profited greatly when Mears's Rio Grande Southern Railroad opened the Rico and Telluride markets. It still endured shortages of flux, ore, and occasionally working capital. Profits rolled in until the second half of 1893, when the crash and mining decline inevitably took its toll and put the company in the red. In 1895 the Omaha and Grant Smelting and Refining Company of Denver leased the plant. Silverton, at this point, offered stiff competition. The hard times worsened when a Durango fire destroyed several buildings that year. Weathering this storm, the smelter, in April 1899, merged into the American Smelting and Refining Company, a trust which went on to dominate American smelting. Shocked, but not awed, by this development, Colorado employees struck for an eight-hour day at the same pay as the old ten- to twelve-hour day, thereby closing smelters throughout the state from June to mid-August. Smelter owners had countered with an hourly wage proposal at the old rate. The editor of the *Durango Evening Herald* at first supported the men, commenting on June 3 that they had the right to organize and the right to an eight-hour day. The strike dragged on, affecting Durango's coal mines and businesses, before the union simply called it off. The editor breathed a sigh of relief and hoped that this would be the last labor trouble for years. The changes of 1899 came too swiftly to be digested; the next few years would show what they meant.

One other smelter operated in Durango, the Standard Smelting and Refining, located about a

Drilling by hand into the heart of a San Juan mountain was slow, hard work. That is why the power drill was such a wonderful innovation, even though some miners did initially resist it. The Camp Bird mine was one of the most modern of the turn-of-the-century operations.

John Porter's smelter in Durango was the San Juan's largest by the 1890s. Smelter smoke polluted the air in many communities in these years.

half-mile south of the San Juan Smelter. Specializing in copper-based ores, it also fell under the control of Omaha and Grant and joined the American Smelting conglomerate in 1899. Opening in 1892, it based its anticipated profits on Red Mountain ores, an expectation crippled by Silverton's smelter and the failure of the mines.

The mine owner cast a wary eye on the smelters, regardless of where they were located. His profits depended upon the amount smeltermen would pay for his ore. For example, Otto Mears signed two contracts, one based on 95 percent of the New York silver price, plus a penalty of 50¢ per ton for each percent of zinc in excess of 10 percent, and the other based on the percentage of lead, at a flat $3 or $7 smelting fee per ton. Rickard called it "internecine conflict," claiming that, while he was manager, the smelter people imposed the highest rates traffic would bear and he was saved only by the fact that his ore was needed as flux.[8] The struggle went on, with the arrival of the trust tipping scales in favor of the smelters.

As Durango and its smelters grew and prospered, so did Durango coal mining; the reverse held true as well. The Porter Fuel Company, organized in 1890 by John Porter, who seemingly dabbled in everything, was the largest. It owned coal lands from Durango to Hesperus, and the opening of the Rio Grande Southern not only tapped its tipples and promoted the town of Porter, but also opened a tremendous market in Dolores, San Miguel, and Ouray counties. Throughout the San Juans, coal replaced wood as fuel, and Durango's coal was the most reasonably priced. Production totals trailed far behind eastern slope mines and would continue to do so, though the output increased. For local consumption, however, Durango's product more than held its own as a "clean" coal. The coking veins around Durango also were rated as Colorado's finest.[9]

In the nineties hardrock and coal mines and smelters all fell under company control, frequently absentee directed. As surely as this happened, unionism secured a stronger hold on employees. Impersonalism replaced earlier fraternization, an hourly job replaced unlimited individual opportunity, and individual worth seemed less important than company profits. Conditions like these spelled opportunity for the union organizer. The Western Federation of Miners (WFM), organized in 1893, moved quickly into the San Juans, building a base upon some local unions already there at Red Mountain, Creede, Ouray, Rico, and Telluride. In fact, at the WFM organization meeting in Butte, Montana, in May 1893, three of the four Colorado unions represented came from Creede, Ouray, and Rico.

The goals of the Western Federation appeared harmless on paper. They did not wish to create trouble or advocate strikes; they wanted just compensation for labor, "fully compatible" with the dangers of employment, wise mining legislation, and the right to use earnings free from the dictates of any person. The organizing notice in the *Rico News*, though, called upon miners "to protect yourselves," and do your duty to your wives and little ones by uniting. It further warned not "to deceive ourselves with the belief" that the present standard of living can be maintained without a struggle.

When it became obvious that the local unions did not possess enough clout individually, they combined in September 1898 to form a district union. The preamble to its constitution stated: "Men engaged in the hazardous and unhealthy occupation of mining should receive fair compensation for their labor and such protection from the law as will remove needless risk to life and health." Although promising to use all "honorable means" to maintain friendly relations with employers, the constitution carefully described local strike procedures, including a secret ballot, a three-fourths resident membership yes vote, and a 15-day notice to the company. The battle lines were being drawn, the whirlwind reaped.

The first major labor conflict to hit the San Juans came in 1893, a reaction to a general attempt to lower wages to $3 per day. Lower silver prices were blamed, a convenient scapegoat. Walkouts hit Rico, Summitville, and Telluride; however, management held the upper hand. These local unions were not financially strong, and the times were such that replacement miners could easily be found. In the end the men lost; the one factor that impressed them

One San Juaner whose mine failed did not seem to begrudge his loss: "Well, I can't help laughing when I think of what a d----d good thing we would have had if we had only struck it." He would have loved the Silver Lake mine, whose mill is shown here down in the Animas Canyon.

was the need for a stronger and larger union.

Scattered strikes followed, such as the ones that hit the American Nettie over the price of board, the Guston in 1895 regarding wages, and the Revenue Tunnel in 1896-1897, involving hiring a shift boss and wages. Beneath the surface calm, the situation seethed. The companies disliked the WFM and worked to undermine it wherever possible. The Revenue trouble (the Virginius Mine, basically) showed what could happen. The company promptly shut down in December, then reopened in January 1897 with a new crew. An attempt to bring in zinc miners from Missouri was repulsed by strikers, while the union urged caution, peace, and order. In the end, the company's strength won out and many of the old crew were not rehired.

Wages in the San Juans became a volatile issue. The range for miners ran from $2.50 to $3 per day in 1897, lower than earlier. Stoiber paid his foreman $4.50 to $6 per day, machine men (power drill) $3.50, and shovelers $2.75. Some attempt was made to introduce the contract system, by which miners were paid so much per square fathom of excavation and could theoretically make more than the standard wage. The plan was not instituted universally, nor was it popular with the miners, who claimed management weighed the odds so that a man could hardly dig enough to make $3 per day, let alone more.

Another festering issue was the appearance of "cheap and foreign labor," in this case from eastern Europe. In 1891, Guston miners at Red Mountain complained that foreigners on contract were to be shipped in but offered no proof. A glimpse of events to come occurred at Henson, nestled beside the Ute and Ulay mine. When the company required that single men board at the company boarding houses in 1899, Italians working there refused to comply and marched out, but Americans tried to continue and were driven away. When Springfield rifles were stolen from the local armory and armed Italians seized the mine, the local sheriff called for Colorado National Guard troops. Ill will toward the Italians mounted and the troops arrived just in time to calm the situation in mid-March, though not before the local press cried the "Italians must go." As a result of the uproar, the company agreed not to hire any more Italians, an injunction was issued to restrain the strikers, and the Italians were given a few days to leave the district.[10]

The use of troops, the injunction, and racial tension were not new to Colorado, but for the San Juans they heralded a bitter day. As disquieting as management-labor confrontation had been here in the 1890s, it had been more convulsive elsewhere. The Western Federation called major strikes at Cripple Creek and Leadville, giving the owners as good as they gave. Both sides dug in, neither willing to compromise and each growing to hate the other. The issue of union recognition was guaranteed to generate sparks, management being doggedly determined to wrest the initiative from a union which it conceived to be a threat to the very fabric of America, the free enterprise system.

The union held out promise of a better deal to the miner. For a while, the state did also, when a union-backed eight-hour day bill became law, only to have the Colorado Supreme Court declare it unconstitutional. Though the act died, the issue did not. For the miner working his shift, promises were small comfort; his job was still physical, dangerous, and dirty, even with the mechanical improvements. Premature blasts, falls, human error, a score of possibilities kept his work dangerous; many miners carefully weighed marriage, because insurance was virtually unavailable or too costly, leaving his family destitute if a fatality should occur. Those new drills spewing forth clouds of razor-sharp rock dust were not nicknamed "widow makers" out of love. Low pay and hard work were the miners' lot. As a result, some miners high-graded (pocketed rich pieces of ore) to supplement their pay, a practice which drove management into a frenzy to stop. Rigid rules and searches slowed high grading; also effective was firing, a policy which resulted in the entire crew of the Golden Fleece being booted in June 1895.

As the decade closed, the San Juans had progressed notably since 1889. Not all these advances came without some waste and mismanagement. A geological report of the period was critical: "Capital thus invested serves only to build monuments of failure, folly and dishonesty, which may operate to delay for years such development and improvement as the mineral resources of a district really warrant." Fortunately, less waste occurred than previously. Chauncey Hall summarized San Juan developments better when he rhapsodized, "A well developed and prudently managed precious metal mine is one of the grandest pieces of property in the world."[11] Silver had once been queen, now gold reigned, a golden monarch that promised to carry the San Juans to grander heights in the twentieth century.

[1] J.S. Robinson to Henry Moody, March 13, 1892, Henry Moody Collection, Western Historical Collections, University of Colorado. For Creede, see *Engineering and Mining Journal*, 1890-1894; *Financial and Mining Record*, 1891-1892; *Creede Candle*, 1892-1894; William Emmons and Esper Larson, *Geology and Ore Deposits of the Creede District, Colorado* (Washington: Government Printing Office, 1923).

[2] *Financial and Mining Record*, July 4, 1891, page 7.

[3] For Silver Lake, see Edward Stoiber, "General Report Upon the Silver Lake Mines, 1899"; Silver Lake Mining Records, San Juan County Historical Society. For the Tomboy, see *Engineering and Mining Journal*, 1891-1899; James D. Hague Collection, Henry E. Huntington Library. For the Smuggler-Union, see *Engineering and Mining Journal*, 1890-1899; John Porter Letters, William A. Bell Papers, Colorado Historical Society; *Telluride Journal*, December 30, 1899.

[4] Helen Croft, *The Downs, The Rockies and Desert Gold* (Caldwell: Caxton, 1961), page 56.

[5] Sources on the San Juans for the 1890s are numerous. The *Engineering and Mining Journal, Financial and Mining Record*, and local newspapers were used. See also Otto Mears Papers, Colorado Historical Society; Buchanan and Searcy Collections; George Roberts, *Report of the Director of the Mint* (Washington: Government Printing Office, 1898, 1900); and J.S. Robinson to Moody, March 1, 1892, Moody Collection.

[6] J.S. Robinson to Moody, February 11, 1892, Moody Collection. Stephen Bailey, *L.L. Nunn a Memoir* (Ithaca: Cayuga Press, 1933), pages 60-77. *Financial and Mining Record*, July 4, 1891, page 8. Irving Hale, "Electric Mining in the Rocky Mountain Region," *Transactions of AIME* (New York: AIME, 1897), pages 403-412. *Engineering and Mining Journal*, 1891-1896. Tomboy Notebooks I, Hague Collection.

[7] John Farish to Isaac Ellwood, May 20, June 5, and August 22, 1891, Isaac L. Ellwood Collection, American Heritage Center, University of Wyoming. T.A. Rickard, *Retrospect an Autobiography* (New York: McGraw-Hill, 1937), pages 53-65. See also Herbert Hoover, *The Memoirs of Herbert Hoover: Years of Adventure 1874-1920* (New York: MacMillan, 1951), pages 25-34.

[8] Smelting and milling sources: Rickard, *Retrospect*, pages 64-65. Mears Papers. *Liberty Bell Gold Mining Company Annual Report* (Kansas City: n.p., 1900), page 12. Roberts, *Report of Director of the Mint for 1897*, pages 118-121. Porter Letters, Bell Papers. The San Juan Smelting and Mining Company Annual Reports, 1890-1894, Bell Papers. *Engineering and Mining Journal* and local papers, 1890-1899.

[9] *Durango Land and Coal Company Annual Reports* (Colorado Springs: Gazette Printing Company, 1890, 1893, and 1894). *First Report of the Porter Coal Company* (Denver: C.J. Kelly, 1891). J.S. Robinson to Moody, March 3, 1892, Moody Collection. *Engineering and Mining Journal*, June 24, 1899, page 748. Durango newspapers.

[10] Sources on unions and labor unrest: Archives of the Western Federation of Miners, Western Historical Collections, University of Colorado. *Engineering and Mining Journal*, 1892-1897. Local newspapers. Roberts, *Report of the Director of the Mint for 1897*, pages 118-121. Silver Lake Mine Paybook 1898, Center of Southwest Studies. *A Report on Labor Disturbances in the State of Colorado, from 1880 to 1904 Inclusive* (Washington: Government Printing Office, 1905), pages 102-105. Richard Lingenfelter, *The Hardrock Miners* (Berkeley: University of California Press, 1974), pages 132, 164, 194, and 216-217.

[11] C.H. Hall Letter, *Financial and Mining Reporter*, October 24, 1891, page 295. Frederick Ransome, *A Report on the Economic Geology of the Silverton Quadrangle, Colorado* (Washington: Government Printing Office, 1901), page 21.

T he "Gay Nineties," those years romantically portrayed as a simpler life, displayed a darker side to the San Juans. With their economic disruptions, political crusades, and emotional elections, they would be long remembered though rarely savored. Within these traumatic years some progress was made. This should be remembered when one reads Lucien Nunn's bleak 1889 analysis; he wrote a friend that "on the whole this country affords few inducements" to a young man who had any other object than success by the strictest industry, economy, and "I might almost say privation." The temptations of dissipation, he concluded, were numerous and entertainments few.

Minnie Young's Piano and Kindred Matters

More changes took place in the 1890s than the San Juaner had ever beheld before. The very electricity Nunn experimented with altered the life style significantly. Privation, in the sense of pioneering privation, was finally obliterated from the entire region with the completion of the railroad system. The temptations persisted, curbed somewhat by public pressure; a mining community could not abandon its heritage that quickly. However, Nunn's desired entertainments multiplied.

San Juan urbanization reached its zenith during the nineties, each of the towns surrounded by satellite camps. Gone now were the rush, the rawness, and the restlessness of the birth and the first boom; in their places came stabilization and a more routine (as applied to a mining community) day-to-day life. To the forefront marched those fortunate towns with a favorable geographic site, railroad connections, the county seat, strong civic leadership, productive tributary mines, and a strong confidence in their community's future. To the rear slipped those camps not blessed with the right combination of these attributes. Both trends became evident to resident and visitor; no more could boosterism and optimism mask the obvious. Some camps had already disappeared, as others slid rapidly toward oblivion.

There is always an exception to generalizations. That exception was Creede, a camp with a style the San Juaners had not observed locally. Not that what happened there had not happened before; it was just that in this place it was compressed into a few short years with abundant publicity. Creede materialized in a twinkling, commanding the elements that the public thought to be indigenous to the mining camp—glamour, excitement, and wealth. But there was more to its attraction. There seemed to be a foreboding that Creede was the last fling, that the "mining west frontier" was passing, and that to experience it before it was too late one must go to Creede. The *Rocky Mountain News*, January 19, 1892, expressed the feeling: "It is the only place of the true frontier type in Colorado and it is doing its best to keep up its end. A year from now it will have changed . . . ." Thus they came to visit and to stay, finding, as Nicholas Creede had forecast, that "some will make money and some will go broke."

Other new camps appeared, a few near Creede, but none could hold a candle to it. Creede's eminence was evanescent; as one viewer described it, "More like a circus-tent, which has sprung up over night and which may be removed on the morrow, than a town, and you cannot but feel that the people about are part of the show." A fire gutted the community in June

1892, and before it recovered from that, the crash of 1893 doomed it like "an ambitious youth made old in a day."[1] Creede retrenched then, settling for a position in the second rank of San Juan towns, behind Telluride, Silverton, and Ouray. For a moment in time it had claimed attention, gaining a measure of notoriety for the San Juans that fate had earlier withheld from them.

Creede was not, of course, the only community that felt the effects of the 1893 crash and depression; it was simply the most obvious one. No San Juan community escaped that catastrophe and the burning silver issue. Creede went so far in July as to hold a mass meeting in which participants resolved that the country needed two separate governments, one for the East and one for the West. Such impetuous statements received more national attention than they deserved, but the problem was serious. The closing of the mines and the laying off of miners hurt the economy and affected everybody. Newspapers advised keeping a stiff upper lip, muddling through, and awaiting free silver. The *Creede Candle*, tongue-in-cheek, advocated adopting a gold camp to show faith. The *Silverton Standard* pointed out to its readers that, as hard as times were, "The people of the western states are living on luxuries compared to eastern people." The editor pictured starving working men in the large cities, which provided small comfort to a local Silverton barber who lost faith in banks (the failure rate was startling), withdrew his money, and was promptly robbed.

The months passed and the situation failed to improve materially. Coloradans grew more resentful over what they considered the government's insensitivity to restoration of silver. San Juaners, like the rest, and more emotionally than logically, pinned their hopes on silver. Analyzing this phenomenon, an observer pointed out, "They will not accept the situation, and seem to believe that at some ratio yet to be determined, the free coinage of silver is assured and pending the arrival of that time they are not disposed to do any business that can be avoided." For years San Juaners had been listening to speakers like ex-senator Nathaniel Hill, who told them the United States had been a bimetal country since its founding and that silver had made the country strong. Or Colorado's silver spokesman, Henry Teller, who warned that America's future rested with the people, not Wall Street. The people favored silver, in contrast to Wall Street and foreign bankers, who championed gold. Thus emerged the sinister plot to demonetize silver and confiscate the "reward of the hard toil of the masses," said one-time San Juan visitor John P. Jones, now a United States Senator. One could have bet his last silver dollar that the San Juaners were not about to abandon silver.

From 1893 through the remainder of the decade, a politician's stand or a party's platform on silver became the touchstone for guiding San Juan voters. The issue reached a climax in the 1896 election with the emergence of silver's champion, William Jennings Bryan, the Democratic nominee for president. "Bryan is our man, first, last and always." He would unquestionably lead the way to the promised land of silver. James Hague, in Telluride to examine the Tomboy, tried to describe the excitement to his wife: "In Colorado Bryan is regarded as a Moses, a divinely appointed leader of the people, or, at least, a Lincoln, raised up to save and redeem his people." Teller, the only politician rivaling him in popularity, marched out of the Republican convention when the party refused to support silver and returned to Colorado a hero. Two hundred Rico citizens signed a telegram offering their profound gratitude and thanks "for your loyalty to silver and the common people everywhere." When Teller toured the San Juans in August on Bryan's behalf, he received a heartfelt and elaborate reception. Throughout the campaign money was collected, clubs were organized, and articles printed.

Then came the November decision. Ouray backed Bryan better than 100 to one, Mineral County reported no votes for his opponent, William McKinley, and La Plata's total was 2,764 to 85. That overwhelming local support could not prevent Bryan's and silver's national loss. The *Ouray Herald* consoled: "We are defeated yet not discouraged." Four years hence the banner would be raised once more, and Bryan would again be chosen to "lead the masses to victory."[2] It was never to be. By then times had improved, gold had replaced silver in the San Juans, and fresh issues had emerged. The San Juans loyally voted for Bryan, though less overwhelmingly, only to taste defeat again.

In retrospect, the silver issue was not so crucial for the San Juans as it appeared to be in 1896. For some districts and camps it marked the end of the trail; however, they would have failed anyway, their ore values being low and prospects dim. Politically, it ruined the Republican party in the San Juans for ten years and gave strength to the People's Party, better known as Populists, which promised reform and, most critically, backed silver. For a while the Populists rode high, in 1892 electing five of six San Juan seats in the Colorado House. Eventually, when the furor subsided, they peaked, declined, and merged back into the older parties. Local issues came and went, always overshadowed by silver. Because of the uncertainty of the times and changing political fortunes, incumbents found it difficult to retain their seats. In the 1890, 1892, and 1894 elections, for example, the San Juaners sent totally new representatives to the Colorado House.

The winds of political change helped women finally win equal suffrage. San Juan males decisively approved the constitutional amendment, except in San Miguel and La Plata counties, which stubbornly voted it down. One concerned male predicted "the advent of women is going to change things rather materially"; to what outcome he did not hazard a guess. A year later, the editor of the *Silverton Weekly Miner*, November 2, 1894, advised women to make every effort to inform themselves on the "economic, political and ethical questions of the day," as they prepared to exercise their new right. He recommended a nonpartisan organized society as the appropriate avenue.

No previous experience could chart the changing times. The first suffrage impact was felt locally, though it was hardly an instant phenomenon. Women had been trying with steady success since the 1870s to curb the excesses of their communities. Now they had acquired a forceful weapon, the ballot. Prohibition seemed to be within their grasp, or at least local option. Never a popular issue with an element of San Juan males, the fight against "demon rum" nevertheless accelerated. Finally, a horrified Ouray editor could restrain himself no longer and warned his readers that "prohibition kills any and every community" it masters. If the ladies of Ouray "brought about prohibition, ended gambling, closed dance halls and houses of ill repute, why half the people and business interests would depart, leaving a dead town in their wake." Not so, shouted the ladies, and the battle continued deadlocked.[3]

Women likewise made their presence felt at city hall, where ordinances were on the books restraining gambling, dance halls, and prostitutes. Strengthening of current laws and enforcement of them became the issues. In 1895 Lake City approved an antigambling ordinance and backed it up with stiff fines. So did Rico, although the fines were less, and they also passed a detailed ordinance about "offenses against public morality." Following public discussion, Ouray's council agreed to a children's curfew, setting 9:00 p.m. as the magic hour after which they had to be accompanied by their parents. Rico more rigidly enforced its ordinance pertaining to minors in saloons and gambling halls. Creede, of course, started from scratch, passing the usual regulations involving the "social evils." Because of the times, enforcement vacillated until past mid-decade. One sidelight to Creede's council problems was public dismay over a dirty city hall, resulting in a thorough cleaning of the premises. Not all efforts produced results—an attempt to enforce the Telluride ordinance regarding women loitering about saloons was thwarted. Women never fought alone; they always had strong support from some of the men. A Telluride livery and feed business operator, for instance, complained to the board that the woman next door was conducting a house of prostitution. He claimed the establishment was a menace to the safety of his property and reminded them that they had regulated the places where prostitutes could operate, which did not include the territory in which his stable was located. After a discussion, the board concurred and the

(top) Not all mining camps accepted religion. The Rev. William Davis was convinced that if Red Mountain had helped him build a church, God would not have devastated the camp with the 1892 fire. Rico, however, was one of the majority that did have a church. This one, built in 1892, evolved from a Congregational to a Presbyterian to a community church.

(bottom) The baseball team and town band were sources of local pride. Telluride's cornet band is going somewhere on an excursion. The Rio Grande Southern engines and trains were new in this early 1890s trip.

house was ordered closed.

Municipal governments confronted a variety of challenges during the decade; their responses varied, depending on the presumed immediacy of the issue and the community's current circumstances. As an example, the crash of 1893 could hardly have been ignored, but the minutes of the Telluride and Rico meetings leave the reader with the impression that the city fathers felt no urgency. A few items merely hint at the issue: Rico's register of business licenses showed the number issued dropped nearly two-thirds by 1894; the water works committee recommended no action on the proposed new works because of the "present financial stringency"; and in April 1894, a special canvass of businessmen raised money to help pay the night marshal's salary. Ouray and Creede both lost money in banks that failed, and eight officials of the latter municipality moved away in six months, necessitating wholesale changes. Ouray forthrightly responded to the emergency by allowing smaller payments on business licenses, reducing municipal salaries by 20 percent, stopping some city services (street sprinkling, for example), and asking the electric company for a reduction in price or a cut in services.

Each of these four faced problems with its water system. Growth and its attendant demands for more and better service forced expansion or complete rebuilding, leading to the sale of water bonds and higher license fees to help pay expenses. Customer nonpayment generated much discussion and dismay, as did concern over payment of the bonds and repair of the system. The public, well aware of water problems, voiced opinions without hesitation in meetings and in print. Sanitation troubles persisted, most noticeably in the growing communities. Even with ordinances and public and private collectors, the mess did not markedly improve. In fact, as ordinances against animals at large were enforced, such natural scavengers as pigs and burros were removed. Practicality ran headlong up against civic maturity, with no acceptable solution. Enforcement of a dog license tax also tried the city fathers' patience. "Fido," "Tige," "Paddy," "Browney," and "Rover" roamed the streets at will, as authorities were unable to corral or even devise a means to control them. Thus for another decade the dogs won out.

Some strides were made in raising revenue, a mill levy now generally being assessed on the valuation of taxable property, supplemented by business licenses and other sources. In Creede's case the police judge was allowed to keep 10 percent of the fines collected from bawdy and gambling houses, padding his income considerably. Community advertising did not slacken; a special effort was made in 1893 to have Creede represented at the World's Fair in Chicago. Hard times or not, it could not afford to let that kind of opportunity pass.

Civic maturity and women's increased influence, supplemented by a broadening economic tax base (no longer was the sub rosa revenue from fines so badly needed), generated pressure to curb the excesses of the red-light district. Under constant attack and losing its economic importance, the red-light district was ripe for regulation. Because of its newness Creede showed less of this tendency toward respectability; the others struggled to turn the corner. Complaints and petitions mounted, tentative steps were taken, and then came actual enforcement. Rico took an unusual step by stating that bawdyhouse keepers would be responsible for any disorderly conduct in their establishments. If this rule was not obeyed, then the ordinance prohibiting such houses would be enforced. Ineffective enforcement or practicality, perhaps both, dictated this course. Even Telluride, with its record of lax enforcement of ordinances, prohibited women from drinking in public and prostitutes from loitering. It went so far as to ban a lady pianist from a saloon. What happened by decade's end was that those elements deemed undesirable were either pushed into defined sections or out of town. This was a nineteenth-century version of zoning legislation.

Foul water and a sullied environment concerned San Juaners, who could find no easy remedy. The Telluride board of trustees questioned the dumping of garbage into the San Miguel River in 1897 and finally ordered it stopped and removed, most emphatically the dead hogs. Rico

The crowd gathers at Eureka to see the first Silverton Northern train in 1896. The tracks later ran on to Animas Forks. Eureka's hotel was not imposing but is a classic example of a log, false-fronted building.

gave its marshal the same task, as if he did not already have enough to do. This was typical; the marshal was the official on whom the most demands were made, and the one whose salary caused the most haggling. Demonstrating an awareness of safety, peace, and quiet, Creede banned fireworks within the city limits as early as 1893. Enforcing this ordinance was undoubtedly another matter.[4]

The towns appeared to be better governed and regulated than in the eighties. More cognizance of municipal matters and problems was certainly being shown by both residents and their elected officials, not that the latter rose above criticism. Dave Day blasted Ouray's leaders one Easter time: "We trust the mayor, city council and officials will attend in a body [Easter services] as it is idle to attempt to conceal the fact that they are sadly in need of advice of a sacred and elevating nature." Charges of corruption and misgovernment flew, usually around election time. In the end, however, it was one's neighbor who served on the council or as an official, and both voter and representative stood publicly accountable in these relatively small communities. Businessmen retained a strong hand on power. Creede's government in 1892 consisted of a lawyer, banker, hotelman, lumberman, saloon owner, freighter, storekeeper, and mine owner; several of the others also listed mines, but not as a primary occupation.

Young municipal governments wrestled with problems similar to those their older neighbors had encountered a few years before. Bachelor, a small camp above Creede, elected its first board of trustees which, in the July 16, 1892 meeting, rushed through a contract for a jail, passed sixteen ordinances, and handled a variety of other pressing topics. The agenda soon slackened and urgency decreased, as it always did, and by September a regular routine had replaced those early pressure-packed meetings. The approved ordinances paralleled those of Silverton or Lake City in their youth. Bachelor, alas, was not destined for long life; much of what its board seriously debated and discussed proved moot, for within a decade the ordinances, and even the board, were not needed.[5]

San Juaners found themselves virtually in the mainstream of national developments. News that once had taken days to arrive now flashed over the wires within hours. Instant communication benefited businessmen and law enforcement agents, among others, and heightened public awareness of state, national, and even international events. Gone was the isolation barrier. When the United States declared war against Spain in 1898, Rico decorated her streets and public buildings within a day and offered a volunteer company to the governor. San Juan newspapers carried current reports from Washington, and only the most secluded miner did not know about the war within a week. San Juaners marched readily to war with a boomingly cheerful air; "Let 'er sizzle," crowed the *San Miguel Examiner*. There was some concern that the Indians might start trouble in the absence of regular troops. Showing the old western attitude, the same paper stated that, should this happen, there would be "plenty of 'good Indians' " and no need for regulars after the war ended. Another editor commented wistfully that the war might result in the free coinage of silver. Excitement soon waned, military action dwindled, and the war column disappeared. The San Juans returned to mining, celebrated the glorious Fourth with the usual high spirits, and watched the short war draw to an end. Peace discussions proved less newsworthy and, like many of their contemporaries, the San Juaners scarcely appreciated the implications of this international milestone.

When the bicycle craze hit America in the 1890s, San Juaners pedaled and pushed those new safety bikes to unlikely places over roads ill-suited for those vehicles. They joined the League of American Wheelmen and supported better county roads. A minister was spotted riding to church, and the bicycle drummer became a harbinger of spring. Bicycles were not cheap; the Monarch sold from $40 to $100, "perfection" the company motto. For twenty cents, the Monarch people offered a deck of playing cards that featured that belle of the nineties, Lillian Russell, figuring they would enliven a San Juan card game and promote their product at the same time. Thoughtless riders threatened pedestrians, and the Silverton city fathers instructed the marshal to keep riders off the sidewalks.

Creede's marshal faced a different problem. Without doubt this town had as well organized an underworld as any of the large eastern cities; Jefferson Randolph "Soapy" Smith saw to that. A shrewd natural con man, he hurried down from Denver to take over the sporting element and control the red-light district. Bob Ford, of Jesse James fame, opposed him for a while, then grudgingly submitted. Smith managed to walk a tight line among Creede's various elements and retained power until the crash of 1893 ended the free-wheeling days, prompting Soapy to drift on to greener pastures. No other San Juan community beheld the likes of Soapy Smith. Of course, he was not openly lawless personally. He agreed fully with nearby Bachelor's desire to avoid a reputation as a tough town. A skillful manipulator, he had his loyal supporters elected or appointed to office. Smith parlayed the unsettled situation and the permissiveness of the residents into personal power. As long as nothing got too far out of hand, he was able to build his little empire.[6] Creede's sudden growth and transitory populace played neatly into his schemes, a game which the times and concerned residents eventually closed.

The day of the rough-and-tumble frontier lawman rapidly receded with the arrival of the railroads, women, and "down east" civilization, but one of them still held out at Telluride.

Columbines were gathered up by the armload and shipped to Denver, as well as gracing local homes. These folks in front of Howardsville's general store are nearly lost behind the display. Mining people threatened the environment in more ways than just mining.

In 1892, as in other winters, it was important to have hay to feed the animals. If a Telluridian wanted his horse shoed or needed to rent a sleigh for a moonlight ride with his sweetheart, he had only to go to one of the two establishments to the right.

He was ex-Quantrill raider from the Civil War's border warfare and sometime outlaw Jim Clark, now a town marshal. This large, brown-eyed man strode directly out of a legend. The "greatest fighting man" he ever saw, claimed a contemporary. Another said, "A great fighting man, and physically a strong man; in fact, a real fighter with a gun or any other way." And, finally, the *Rocky Mountain News* believed Clark took more pride in his revolvers and his horse than in anything else. Gunnison County sheriff Cyrus Shores called his fellow lawman a "very strong and active man," who "always appeared to act in a dignified kind of way," one who kept things quiet, "no noisy demonstrations made round town by anybody drunk or sober, that lasted for long at any time." Clark's reputation made his work easier.

There was nothing simple about this man who did not drink (one of his favorite expressions was "Old John Barleycorn will beat them every time"), despised prostitutes, treated children kindly, generously helped those he felt to be in need, and kept peace in his own way. His virtues did not preclude his taking a cut from robberies in his district, including, he confessed to Shores, money from the previously discussed Telluride bank robbery. Finally wearing out his welcome, Clark was dismissed by the council and left office a bitter man, threatening to kill the members at 15¢ each or two for 25¢. Considering his disposition, this was probably not an idle threat. Clark did not live long enough to carry it out, in any case; an unknown midnight assassin killed him from ambush on August 6, 1895. Fifty-four when he died, Jim Clark symbolized both the best and the worst of an era now almost gone, except in the fictionalized "gunsmoke and gallop" West.

The new and the old merged in the San Juans. Even though the Indian issue had long since been resolved locally, an underlying hostility remained, as the previous editorial comment indicated. Bigotry toward the Chinese persisted, occasionally breaking into violence, and a nagging fear of the eastern European immigrant bubbled to the surface in the late nineties. The blacks, too, felt discrimination's sting. In Rico a boxing crowd found the referee so flagrantly prejudiced against the black boxer that it demanded and got a new arbitrator. A Durango newspaper commented on an attempt to pass resolutions against "jim crow" railroad cars, lynchings, and the use of the word colored:

> They [black Georgia newspaper editors] may, in the course of a few centuries, get the 'jim crow' car abolished and diminish lynching; but they will never on this earth get the people of the south to call a 'niggah' a 'negro.'[7]

It was still deemed important to enhance a community's image with a church. Methodist Bishop Henry Warren came to Telluride to raise money for his denomination. He called at the office of the Tomboy Mining Company, where he encountered manager John Herron. The highly respected Herron, born of missionary parents, claimed he had grown "somewhat rusty" since coming west. With a twinkle, the Bishop countered, "Mr. Herron, the gold in the Tomboy Mine has been there hundreds of years, but has not grown rusty." Warren received a $500 donation and, before departing for Denver, secured money for construction. With a two-decade toehold in the San Juans, the churches faced the continuing problems of money and membership. Women remained their primary support. Finances plagued the members in the nineties, as they had previously. The "tainted money" of saloon keepers and others often rode to the rescue of a beleaguered minister or church whose members would not welcome the donors for Sunday services.

In an age which uses the telephone as a matter of course, it is hard to imagine a time when it was considered a newfangled contraption. Battered by storms and repair bills, telephone lines first appeared in the San Juans in the eighties and gained a firm hold in the nineties. To progress from the novelty of a Sunday afternoon concert, when various party line users would perform a musical selection while the others listened, to the status of being taken for granted took about a decade. For a worried parent, the lonely housewife, and the progressive businessman, it was a godsend; for the entire region, it was a definitive sign of coming of age.

Durangoans also rode on excursion trains to follow their baseball team or go to another community's July Fourth or Labor Day celebration. These young girls and fellows are quite nattily attired for their trip. As clean as they are, the trip must be just beginning.

Creede's street is typical of the crowded, littered condition found in mining communities. Jerry-built, unpainted, and filled with transients, they did not provide the best of living conditions.

Railroad building continued. Otto Mears enthusiastically completed the Rio Grande Southern (RGS) from Durango to Ridgway via Telluride and Ricó, and in 1896 the Silverton Northern to Eureka. Mears had overextended himself this time, however; the Southern went into receivership in 1893, a victim of the hard times. An embittered man finally gave up the fight to save the RGS, and the Denver and Rio Grande, which had been financially involved from the start, acquired complete control in 1895. Mears's Silverton railroad, serving Red Mountain, faced equally trying times, but he was able to retain it. He did curtail operations during the winter to save the large sums spent on snow shoveling. Alexander Anderson, superintendent of the Silverton railroad, commented in July 1898 that "slim pickings" faced the Silverton and complained that business was "dwindling down to nothing." Within a year this line, too, was in receivership.

The Silverton Northern did somewhat better, though even here superintendent Anderson encountered problems. One year the Animas River washed out part of the track bed; another time an engine killed a cow, which the owner promptly categorized as a prize bovine. Freight rate complaints hounded Anderson. Approached in 1898 to join in building a road from Silverton to Gladstone, Mears declined to put more money into the San Juans. The Silverton, Gladstone, and Northerly was built the next year anyway, making Silverton, with four lines, the railroad capital of the San Juans.

Other roads never advanced beyond the planning stage. A reasonable person could see that the San Juans had become well railroaded. All the major towns and districts were tapped, and from the railroad depot freighters and packers pressed on to the more inaccessible locations. The Denver and Rio Grande retained its hold on the region; not even Mears could successfully challenge it. Better financed, larger, well managed, and ruthless, the D&RG held the San Juans with a firm grip. Its awesome power aroused jealousies. Complaints about freight rates, particularly related to ore, indicated dissatisfaction among its customers. Anderson, who considered the D&RG less than fair when it told the public that his Silverton railroad ran very irregularly, grumbled, "I don't mind fair competition, but I think it too bad to deceive the public."[8]

Old timers, however, could remember when there had been no railroad and knew what a difference it had made. They could remember many things about the quarter of a century that had passed since "Ho for the San Juans." They organized a San Juan Pioneers Association for those who had arrived before July 4, 1880 and held annual meetings in various places. One of Telluride's papers ran an "Old Timers' Gossip" column, referring to the ancient 1880s, and some Ricoites bemoaned the fact that no celebration marked that town's twentieth birthday in 1899. Mutterings were even heard that times were not like the "old days." Countering that, one realist pointed out that many had come, as he had in the seventies, expecting to leave with a "trunk full of money." Still in attendance, he honestly observed that the money in his trunk "doesn't make it necessary to sit on the lid in order to lock it."

Speaking of the "good old days," how do the 1890s fit that description? Disregarding for the moment the obvious improvement in living conditions since 1870, it is evident from the files of the Silverton lawyers, Buchanan and Searcy, that those times were often less than "good." A client wanted to know what happened to those men who volunteered to sign his bail bonds, "Why in hell don't they do it." Numerous times the lawyers tried to trace persons who had left town owing money, or to collect money locally. One man wanted his brother to draw his pay check, "so that I have nothing to do with my checks." The lawyers became embroiled in a legal fight between a son and his stepmother over some land and wondered if they could make a jury believe the boy's case. Out-of-town merchants worried about local customers' credit ratings and asked for an investigation. Responding to an inquiry, a correspondent informed Buchanan that the lady in question was much married, at least four times without benefit of divorce. People inquired about jobs, some wanting to move west because of their health; whether Silverton could answer that need was questionable.

One of the more humorous episodes involved a piano owned by a Durango merchant. A music teacher had failed to pay rent on it, and the owner requested that the lawyers sell it. After some dickering, it was finally sold to "Miss" Minnie Young, who was opening a sporting house. There followed a series of letters between lawyers and client about payments and insurance, which the worried owner insisted be placed with a reliable company. Precautions aside, the piano came back to haunt them, when Minnie apparently failed to make payments (was business that bad?). Buchanan and Searcy went back to advertising.[9]

Local merchants also found times hard, and not just because of the poor economic conditions. With luck, drive, knowledge of the market, and a willingness to gamble, the successful ones built up a dependable local clientele, sometimes even a small monopoly. Now this advantage was being eroded by mail order business and the railroad. The railroad meant that merchants in Denver and other large cities were just a ticket away. A note in the *Lake City Times*, October 1, 1896, gave another reason. Quite a number of people were going to Denver the next week to buy goods, at the same time "owing merchants at home" for goods purchased months or a year ago.

The towns still bickered among themselves, jealous of the slightest advantage one might gain over another. Finally, an exasperated editor of the *Ouray Herald*, May 28, 1896, admonished, "In the chase for the almighty dollar even the 4th of July is squabbled over for fear that a neighboring town will get a little the best of it."

People whose childhood recollections centered on the nineties probably had a strong predilection for calling these the good old days. Picnics, baseball games, July Fourth celebrations, drilling contests (gaining rapid popularity), circuses, Christmas holidays, ice skating, outdoor fun—all these were things to be fondly remembered. As they grew older there were dances with the first boy or girl friend, initiation into a fraternal organization, parties, and moonlight sleigh rides. The San Juans offered them all, and who could argue that these times were not, for them, good? They also had better schools and teachers, a fact less likely to impress youngsters.

Teachers, meanwhile, had better toe the mark and measure up to Victorian standards. Dolores County's superintendent of schools refused to certify anyone "addicted to the use of tobacco or any other habit inconsistent with the school law."

Tourists could have been left with a similar glowing impression, as many of them were overwhelmed by the picturesque settings and mountain grandeur. The tourist dollar helped stabilize the town of Ouray and Mears's Silverton railroad, this being the reason that Anderson was so upset over D&RG announcements. The circle route from Denver-Durango-Silverton-Ouray-Denver was heavily promoted. Merchants waited along the way with postcards and other mementos as permanent reminders of that wonderful vacation. When the tourists did not come, concerns mounted. The *Ouray Plaindealer*, June 24, 1898, lamented that tourists were not materializing to any extent. The one group which had just passed through was certainly "good to look upon."

The nineties were an interesting, contrasting, transitional decade for San Juaners. Persistent problems, such as weather, plagued them; one Swede over in Rico reportedly remarked "en summer time et ben vinter time; en vinter time et ben hell." While electricity pointed to a better day coming, the day of the electric kitchen remained a long way off for the hardworking wife perspiring over the wood stove. It had been an energetic, bustling time, marred by the worst regional economic slump yet encountered and a bitter political defeat. A time of growth, a time of decline; a time of progress, a time of stagnation; a decade opening and closing on the upswing, but near hell in between.

Dave Day, still one of the ablest and most readable of the San Juan newspapermen, moved from Ouray to Durango in 1892. A year before he left, he wrote an obituary about an old timer who had crossed "over the range."

Bill and whiskey battled earnestly and agreeably for years, but the Ophir brand [he died in Ophir] was too much for the old man, and he has gone to the land where the streets are said to be paved with the metal he sought and squandered.[10]

The old days and old timers were going; a new breed, who remembered not the trials of those earlier years firsthand, emerged to take their place. The old gave way to the new, and the new would lead the San Juans into the twentieth century.

[1] On Creede see Richard H. Davis, *The West from a Car-Window* (New York: Harper & Brothers, 1892), pages 61-65. Edwin Bennett, *Boom Town Boy* (Chicago: Sage, 1966), *Creede Candle*, 1892-1893. *Rocky Mountain News*, 1892-1893. *Weekly Republican* (Denver), January 4, 1894. The Nunn quote was found in Bailey, *L.L. Nunn*, page 65.

[2] Sources on the 1893 crash and silver campaign: Henry Teller Papers, Colorado Historical Society. George Barclay Moffat to Spencer et al., October 28, 1893, Bell Papers. *Silverton Standard*, 1893, and September to November 1896. Henry Teller, "Discrimination Against Silver," June 10, 1890, John Jones, "Speech," May 12-13, 1890, and N.P. Hill, "Address," August 2, 1893, Henry E. Huntington Library. *Ouray Herald*, 1896. *Rocky Mountain News*, November 5, 1896. *Lake City Times*, September-October 1896. Hague to Mary Hague, November 9, 1896, Hague Collection.

[3] *Silverite-Plaindealer* (Ouray), February 3, 1899. On women's suffrage see R.W. Steele to Thomas Dawson, November 24, 1893, Teller Papers. *Silverton Weekly Miner*, November 2, 1894. *Lake City Times*, November 9, 1893.

[4] Rico, Minutes of the Board of Trustees, 1890-1894 and 1898-1899, Register of Licenses, and Book of Ordinances. Ouray, Minutes of the Board of Trustees, 1890 and 1893-1895. Telluride, City Record, 1892-1894, Minutes of the Board of Trustees, 1897-1898, and *Ordinances of the Town of Telluride* (Telluride: Telluride Republican, 1891). Creede, Minutes of the City Council, 1892-1894, 1896-1897, and 1899, and Book of Ordinances.

[5] For Bachelor see *Teller Topics* (Bachelor), July-September 1892. All the newspapers listed in this chapter's footnotes have comments about local government. Day's comment was found in the *Solid Muldoon*, March 27, 1891.

[6] For the Spanish-American War see: *San Miguel Examiner*, April-September 1898; *Rico News-Sun*, April-July 1898; *Silverton Standard*, July-August 1898; and *Lake City Times*, August and December 1898. The newness of bicycles guaranteed their coverage; for example, *Lake City Times*, May 5, 1892, *Silverton Standard*, June 10, 1893, *Silverton Weekly Miner*, October 19, 1894, *Ouray Herald*, January 30, 1896, and *Rico News-Sun*, April 30, 1898. On Soapy Smith see: Nolie Mumey, *Creede* (Denver: Artcraft Press, 1949), pages 123-135; *Creede Candle*, 1892-1893; *Teller Topics*, 1892.

[7] *Great Southwest* (Durango), December 1892. See also *Rico News-Sun*, March 13, 1897; *Creede Candle*, January 28, 1892; and Evalyn McLean, *Father Struck It Rich* (Boston: Little, Brown & Company, 1936), page 27.

[8] Otto Mears to Hobbs, January 5, 1895, to W. Pitcaithly, April 2, 1896, to Pfaelzer et al., June 22, 1899, and to Barber, January 8, 1900. Alex Anderson to Thorton, June 15, 1897, to McNeil, June 15 and September 16, 1897, June 14, 16, 21 and August 3, 1898, and to Graham, July 18, 1898, Mears Papers. Bennett in *Boom Town Boy*, pages 16-21, has an interesting account of mule skinners. See also Mallory Ferrell, *Silver San Juan: The Rio Grande Southern Railroad* (Boulder: Pruett, 1973).

[9] Letters found in the Searcy and Buchanan Collection, San Juan Historical Society. For the old timers' section see San Juan Pioneer Association Minutes, July 4, 1896, Colorado Historical Society; *Telluride Journal*, December 30, 1899; *San Miguel Examiner*, January 29, 1898; *Ouray Herald*, July 20, 1899; *Rico News-Sun*, August 12, 1899.

[10] *Solid Muldoon*, May 15, 1891. All newspapers cited provide fascinating glimpses into the life and times.

No victory, no liberty," no battle on behalf of the people, can be "won without shedding blood." Thus challenged *The Miners' Magazine*, spokesman of the militant Western Federation of Miners. A eulogy for a dead Telluride member, killed in a fight with scabs and company guards, prompted this:

> But that [his death] is nothing, when we take into consideration the noble purpose, the glorious cause, the honorable battle [July 3, 1901] he was fighting for humanity, when the bullet of a contemptible villain, paid by corporation gold to murder workingmen, found him.

"It is not cheap— lawlessness"

Management responded:

> How can any fair-minded union man contemplate this list of crimes and still endorse this lawless organization [WFM]? How can he approve of an organization that will follow this advice [followers to arm], purchase arms and send armed bodies out to kill and destroy . . .

Feelings ran deep. Following a series of disastrous snowslides, a self-righteous antilabor spokesman declaimed: "I cannot but believe that the recent series of disasters are the judgment of the Almighty God upon the crimes that have been committed in that camp [Telluride] in the name of labor."[1]

While it might be hard to grasp, both sides were talking about the same events, a series of violent labor disputes and strikes that swept the San Juans in 1901-1904. Militant unionism fought entrenched management—no longer with words, no longer simmering, no longer bargaining—now with fists, guns, and dynamite. Across Colorado the battle surged from coal camp to mining town to smelter gate. Damage, destruction, and death were scattered in its wake. The National Guard trooped in, at a conservatively estimated cost of nearly $750,000 to Colorado taxpayers. For what purpose? To decide which faction would dominate.

The chronology and locations may easily be outlined: in the San Juans, they centered around Telluride. The emotions and the significance are less easily discerned. The twentieth century did not give rise to the discontent—the roots went back two decades, nearly erupting in the 1890s. The Western Federation of Miners gained a stronghold in the San Juans by 1901, with locals at Ophir, Henson (Lake City), Ouray, Rico, Durango, Silverton, Ironton (Red Mountain), and Telluride; Creede and Dunton soon joined also. Smarting from earlier setbacks and alarmed by the union's growth, the owners awaited their opportunity. The inevitable occurred in the spring of 1901.

In May the Telluride local called its members out of the Smuggler-Union over management's insistence that the contract or fathom system be used instead of a daily wage. The union charged that, under this system, miners could not earn the $3 daily wages current in the district, and the hurry to mine the fathom (6 ft by 6 ft and as wide as the vein) increased the possibility of accident. Unlike other strikes, this one ostensibly had nothing to do with raising or lowering wages, but focused instead on mining methods. The union offered to submit to arbitration; however, the Smuggler-Union shortsightedly refused and the strike entered its second month. Far too frequently in the months ahead, management displayed similar intransigence. Manager Arthur Collins, an experienced Cornish mining man, then hired nonunion miners at $3 for an eight-hour day, terms denied the union. The WFM countered by trying to induce the "scabs" to quit. Failing in this, they organized approximately 250 strikers and supporters on July 3 and met the night shift as it came off work. John Barthell, a union miner, stood up and shouted for the employed men to come out. Company guards immediately responded, opening fire and killing him instantly. The battle that followed left two more dead, six wounded, and 88 nonunion men captured and brutally driven from the district, an example for which union members would pay many times over before Telluride quieted. Each side blamed the other—small consolation for Barthell and the others, the first victims of Telluride's time of troubles. The traditional July Fourth celebration was dispensed with that year. Governor James Orman calmly refused the demands that quickly went out for troops, sending instead an investigation team.

The outgrowth was a conference in which both sides conceded some minor demands, which eventually led to adoption of the $3 wage and an eight-hour day for men working underground. Though the strike was called off on July 7, bitterness lingered, and the mill and surface workers were not included in the settlement. The heroes or villains, depending on one's point of view, were energetic and successful Vincent St. John, president of the Telluride local, and thirty-three-year-old Collins. Under St. John's direction, Telluride had the strongest local in Colorado, and the union moved to entrench itself throughout the San Juans. It served notice in 1902 that it behooved all miners to join; the owners retaliated in 1903 by organizing the San Juan Mining Association, covering San Juan, Ouray, and San Miguel counties. Both sides jockeyed for superiority, neither listening to the other.

Telluride became a hotbed of activity. The union tried to boycott the vehemently antilabor *Daily Journal*, a move countered by the formation of a Business Men's Association to sustain the paper. According to City Attorney L.C. Kinikin, all businesses "who wouldn't display a 'fair' card in their wondows" also faced union boycott. A hint of trouble surfaced when a wildcat strike hit some of the district's smaller mines in January 1902 over wages. The union gained nothing. The unforgivable incident came when Collins was assassinated at his home on November 19, 1902, leaving a widow and two young sons. A "monstrous crime" committed by an "inhumane, cowardly perpetrator," screamed the *Daily Journal*. This senseless act, denied by the Federation to no avail, convinced the owners that the union would stop at nothing to seize power. That "dangerous agitator," St. John, and eight others were arrested, with no proof of guilt except by association, and the real murderer escaped. For the owners, nothing less than destruction of the union would be a satisfactory solution. In its December 13, 1902 editorial, the conservative, antiunion *Engineering and Mining Journal* darkly hinted that socialism was among the union goals, a real threat to this generation.

In place of Collins came flamboyant, controversial Bulkeley Wells, a charming, handsome, extravagant, and ruthless man, much admired and much hated. In him the owners found a spokesman and a leader. Inexorably, the nightmarish drama marched toward further tragedy. In September 1903 the union struck to secure an eight-hour day for mill men. The managers stood firm, preferring to keep the plants closed rather than "submit to the dictates." The miners marched out. Wells confidently predicted, "The strike here and elsewhere . . . promises to be an absolute failure, and to cause the Federation a large loss of membership and standing."

Martin Wenger remembered the miners coming into Telluride as "a hilarious lot," jamming the saloons and dance halls. Hilarity died as the strike dragged on and spread to Ophir. The WFM opened lodging houses and restaurants on a cooperative plan to sustain members; regardless, many left to seek other employment. Armed union pickets appeared. While the mine owners weighed the use of searchlights and "rapid fire" guns to protect their properties, they began the siege by importing professional gunmen and others to serve as day and night guards. A hard lot they were, including the feared Robert Meldrum, who, according to rumor, killed a dozen men in the line of duty. The county sheriff promptly deputized Meldrum and some of his fellow guards. Meldrum reportedly made his local debut by shouldering up to the bar at the Cosmopolitan Saloon and announcing to the startled patrons,

I'm Bob Meldrum. You can always find me when you want me. Now, if any son of a bitch has anything to say, spit it out; otherwise, I'm going to take a drink—and alone.

Later on, he was daytime guard at the Tomboy mine, where genteel Harriet Backus recalled him as seemingly fearless, looking like the "typical intrepid westerner." The Federation countered with a gunfighter of their own, Joe Corey.

Positions hardened, tension heightened. The crucial issue became survival of the union. Robert Livermore, journeying over from the Camp Bird mine, was shocked to see Pandora nearly an armed camp. Livermore and others tried to get the mills operating with nonunion labor, a scheme that produced a virtual state of war. With some friends, Livermore had to escort workers home, "putting out flankers and driving in pickets" concealed along the roadside.

This was too much and the plea went out for troops. Strongly antilabor Governor James Peabody investigated, then heartily responded on November 20. In marched the National Guard, six companies of infantry and two of cavalry, which deployed at Pandora, Telluride, Ames, and various mines. The overjoyed owners furnished meals and quarters until permanent camps could be established, easing an embarrassing shortage of state funds for such activities. There was no reason that they should not generously support the troops. Every time they had marched in recently they had done so, not as a neutral force to maintain the peace, but on behalf of the owners. The tide had turned at Telluride.

The full force of change soon hit the community. Relentlessly, the owners brought in scabs to replace the strikers, and mining began to resume a more normal pattern. Union leaders were seized for threatening nonunion men, and striking miners, "mostly foreigners," were arrested, charged with vagrancy, and told to go to work, to jail, or to leave. Another group was rounded up and shipped to Montrose, held in jail, and then released. The Federation leadership in Denver responded by protesting this illegal harassment and instituted legal proceedings on behalf of the union members. Peabody's field commander, Major Zeph Hill, concluded that martial law offered the best solution. Once-burned Peabody, however, did not want another court versus military clash; similar action in Cripple Creek had produced harsh criticism of his administration. The mines and mills gradually reactivated and quiet resumed. By the end of December, the troops were being removed. A bleak Christmas passed for the strikers; the mauled local staggered on, in dire straits. Young Martin Wenger thought "things appeared to be normal."

Suddenly, on January 3, 1904, Peabody declared martial law, claiming that the county bordered on absolute insurrection and rebellion. His only justification was based on hearsay and threats of vengeance after the troops left. Hill and the mine owners served as Peabody's informers, capitalizing on their antiunionism to reinforce the determination to eliminate all opposition. They even organized their own troops, volunteered for service, and were mustered in. This amazing declaration coincided with the arrest of twenty-two more union members and sympathizers who stood trial, were deported, and ordered not to return. (The Journal, which survived the boycott, blithely talked of "agitators" being deported and used most of its type on other matters.) Sixty-one more followed the same path that month. Strikers charged with vagrancy were fined; if they lacked funds, they worked off the fine on local projects. When Henri Maki refused to fill a privy vault, he was chained to a telegraph pole for several hours on a cold winter's day, becoming a union martyr and excellent fodder for propaganda. Squads of soldiers patrolled Telluride streets, passes were necessary, a curfew was ordered, freedom of assembly was restricted, and press censorship enforced. No reports were dispatched or telephoned without Hill's approval. A Citizens' Alliance exerted its conservative influence, drawing this scathing denunciation from the *Miners' Magazine*: "no greater aggregation of moral wrecks, libertines, common drunks, diseased both in body and mind," ever polluted Telluride.

Among the guard officers was Captain Bulkeley Wells, who assumed command of the district on February 20. Wenger and his brother obtained passes from Wells to sell newspapers during the curfew, passes they were never required to show "even once." Wenger recalled that the family lost a 22-special rifle, which the soldiers took and never returned when they searched Telluride for weapons. The community was losing much more than rifles—its cohesiveness and solid reputation were rapidly eroding. On March 11 martial law ended. In the meantime, some of the deported men returned. On the night of the fourteenth, they and their supporters, about sixty-five in all, were seized by the Citizens' Alliance, forcibly placed aboard a special train, and ordered to depart permanently. A "nocturnal infamy," "brutal barbarism," cried the union. Worried families and friends again took up the vigil until word was received of a safe arrival at the appointed destination. In the days that followed, a few more union men were deported; there were not many left now. On March 17, Wells was honored with a banquet by grateful friends. The victory seemed to be theirs. The fight was carried into the courts, however, with the Federation losing the first round when it attempted to secure an injunction against deportation. It subsequently won a temporary injunction on March 22.

When some of the deported men announced their intention to return and the Ouray local selected fifty armed men to escort them, Governor Peabody saw "a state of insurrection and rebellion," and back came the troops on March 23, six days after the victory banquet. Captain Bulkeley Wells and Troop A, that group of handpicked Telluride supporters, promptly rode to the rescue, along with martial law. A familiar refrain now was replayed. Civil rights for union members would be forgotten. The town was "much better without the men," thundered the *San Miguel Examiner* on March 19; what was at stake was whether the "mine owners or Union [would] direct management of the mines."

That was how Wells and his backers saw it, and they also saw the battle turn completely in their favor. The same trend moved against the WFM statewide. Deported miners attempting to return were stopped, a few were jailed, and law and order danced to the owners' tune. Martial law and the troops remained into June, concluding a general roundup of "undesirable men in the district." Military rule ended on June 15. Colorado's adjutant general, pompous Sherman Bell, justified this termination by announcing that peace and good order had been fully restored. This was true in the sense that the mines and mills were operating and the owners had acquired complete control. Throughout the summer, occasional attempts by union men to return were repulsed, and "labor agitators" continued to be deported as the need arose. Finally, in November, the decimated Western Federation called off its long-lost strike.[2] A Pyrrhic victory was won when management granted to all employees an eight-hour day, not through negotiation with the WFM, but by unilaterally following the Smuggler-Union's lead. Telluride's time of troubles was over.

No one had wanted it to turn out this way. Not Peabody, on whom the union heaped special scorn: "This poor, weak, miserable, crawling, putty-made sample of effeminate masculinity with a rubber vertebra . . . ." Nor Wells, whom the Federation grudgingly admired as "suave, polite, very much the gentleman." Certainly not Vincent St. John, who had faced guards' fire trying to defuse the July 1901 trouble, then left Telluride, either as a fugitive from justice or in fear of his life, the reason dependent on the source of the story. The other union leaders had not wanted this kind of an end. The Citizens' Alliance did not intend to besmirch Telluride's fair name. City attorney Kinikin said the community was "split wide open." Having defended union clients, he became one of the post-strike victims, losing his position as city attorney. Henri Maki and his fellow miners had had no intention of stirring up such a mess, and the citizens of Colorado had not expected this burden of debt to fall on them. Like a San Juan

The Tomboy mine was one of the richest in the San Juans; production topped $20,000,000 by the 1920s. A settlement grew up around the mine in Savage Basin above Telluride. Unlike many mines, the Tomboy had a talented chronicler, Harriet Fish Backus, who has forever captured the life and times from 1906 to 1910 in her Tomboy Bride. A charming, attractive young lady, she went to the Tomboy about the time this photograph was taken. She and her husband lived at the cabin, Castle Sky High, circled in the center of the photograph.

snowslide, the labor/management dispute gathered speed as it thundered along, going beyond the powers of any individual to control. Obviously, some people were much happier with the outcome than others.

Looking over the startling events of the past months, which included arbitrary arrests, denial of rights, and summary banishments, Ray Stannard Baker, one of the best known of the day's muckraking journalists, lamented, "To this, then, have we come in these American towns at the beginning of the twentieth century! And why is this so? Why have the people borne these appalling usurpations?" Why indeed?

Baker, who was not a union advocate, accused the WFM of being socialistic and blamed the smelter-mining lobby for arrogance and denial of the eight-hour day. He was appalled at the intrusion of a labor problem into politics and the lining up of public officials on one side or the other. The Western Federation, explained Baker, with its unreasonable strikes, violence, and mistakes, was to blame as much as its opponents. The people had broken the law in Colorado; they would have to assume the burdens of maintaining troops , lawsuits, lost business, and rising taxes. "It is not cheap—lawlessness."

With less objectivity and more passion, Bell praised the actions of Wells and others for the "thorough cleaning up of San Miguel County," based on the state power to suppress lawlessness, rebellion, and treason. Wells justified to the Smuggler-Union stockholders what had transpired on the grounds that the WFM had adopted a socialistic program, led by a few unscrupulous agitators. To help stamp out that evil and protect stockholders' interests, he had obligated the company for $18,000 in state certificates of indebtedness, Colorado not having sufficient funds to maintain the troops. It is easy to picture eastern stockholders standing with their representative against such un-Americanism. Both owners and outsiders called the 1901 Smuggler-Union conflict an outrage led by Austrians, Italians, and Slavs. Strong ethnic prejudice against eastern Europeans permeated the strike period. Twenty years later Livermore looked back and blamed it all on the Federation, which ruled with a heavy hand, using terror and murder.

The union had fewer defenders on record, primarily its own publications, but it still managed to present a strong case, accusing the owners of Ku Klux Klan- and Inquisition-type activities in connection with the deportations. Companies were attacked for valuing dividends more highly than laborers, for taking the law into their own hands, for denying human rights, and mostly for trying to kill the union. "Labor," the *Miners' Magazine* defiantly shouted, "is supreme, invincible, unconquerable. The war is on, and it is not going to be settled till it is settled right, which is to be settled in favor of labor."[3]

The war did not turn out that way; no victors emerged and the district and town were indelibly scarred. Management won on paper at the cost of ruthlessly denying labor its rights, intensifying the conflict, and escalating expenses for taxpayer and stockholder. The Western Federation of Miners, its strength broken throughout Colorado, never would have the same power again. By 1909 the Telluride local was "down to rock bottom," with only twenty-three members. Realistically, in view of what happened, it was a miracle that it was alive at all. Colorado lost, too, its state reputation begrimed by the arbitrary, high-handed, antiunion actions of the Peabody administration during the 1903-1904 strikes. That the laborer and miner lost—wages, homes, and hope—is no more clearly shown than by the corporation's complete domination. The rank and file had been ill-used by their leadership to ends not always conducive to the miners' welfare, especially regarding socialistic pronouncements, which had nothing to do with the strike issues. The Western Federation emerged more radical, its leadership socialistic, its membership badly divided on the issue. Those who sought the union's destruction at all costs could enjoy their handiwork; their fear and hatred of the WFM knew few bounds, and on them rests much of the blame.

Baker never really understood why the miners struck, "no miners in the world, perhaps, worked under more favorable conditions and were more generally contented." In a sense this

Winter at the Tomboy, or any mine, meant weeks, if not months, of isolation. Harriet Backus vividly recounts the trials and tribulations of a young wife under such conditions.

The Tomboy's rival across the mountain was the Camp Bird. Both were owned by English companies and each was well equipped and financed. The Camp Bird left nothing to chance, even having its own cobblers to repair hobnail boots.

might have been true, yet these men lashed out because they found their position deteriorating, their world changing, and their jobs becoming impersonalized and dehumanized. They compared their pay to corporation profits and saw inequity. The owners liked to believe that the troublemakers, "socialists," were eastern Europeans out to undermine the American way. A listing of sixty-two deported miners found exactly half from Italy and Australia but the WFM had existed before they arrived in any numbers and had already called several successful strikes. To blame the immigrants was simply an easy scapegoat—the union represented a cross section of all nationalities. The Western Federation gained some concessions for its members, then lost them all in a frenzy of violence. The Telluride local had accomplished a lot of good, including operating a hospital, which the local's depressed condition and heavy debt forced it to deed to the parent union in 1906.

Elsewhere in the San Juans, Ouray and Silverton retained strong locals which physically and financially supported their Telluride brethren. In Silverton, according to a resident, union membership was a prerequisite to working in a mine, and most of the townspeople joined one or another. Silverton also had a miners' union hospital. The Telluride defeat was a blow to them all; Durango, Henson, and Dunton locals folded by 1910, as did the San Juan District Union and its concept of solidarity. Though it garnered most of the attention, the strike at Telluride was not the only one during these years. Coal miners in Durango struck over wages and other problems in 1902, Lake City union members fought a losing struggle in 1903-1904, and in 1905 Alta and Silverton miners contested local issues.[4]

Upon the conclusion of the Telluride strike, Wells announced to the stockholders that henceforth the Smuggler-Union would lease its workings as a solution to the current labor problems. The Tomboy (under John Herron's skillful direction) and the Liberty Bell continued to operate under traditional management. Telluride's "mining trinity" had come back into full glory.

All these mines followed similar patterns, each mining deeper into lower-grade ore. Improvements were made at the mine and mill and in equipment, the Tomboy putting into operation the first electric locomotive in the Telluride district in 1907. The purchase of neighboring claims insured longer mining and stronger legal protection. After weathering the labor crisis, the three went on to prosperous years, the Tomboy production topping $1,000,000 from July 1906 through June 1907; the other two had years over $800,000, for a combined total for the three in the decade of roughly $18,000,000. The prophecy of the seventies had become gospel in the twentieth century.

Wells's leasing policy drew fire. He was accused of giving friends short-term leases on valuable sections of the lode. Even his brother-in-law, Robert Livermore, came to distrust him when Wells mined underneath his lease, stopping Livermore's operation. Wells seemed to attract controversy wherever he went; unconcerned stockholders only waited for those dividends, which he delivered.

In the world of corporation mining, management's role and numbers grew. Alex Botkin worked at the Tomboy as clerk, with responsibilities for issuing stores on requisition, weighing and recording the gold bars sent to Telluride, collecting rents from company houses, and delivering pay checks to the miners. For these duties he was paid $150 per month. With two bosses, one at the mine and the other at the Telluride office, he occasionally found himself torn between conflicting orders. Botkin never had any real mining experience, as he admitted, but he was as much a part of the Tomboy operation as the miner underground, and was better paid.[5]

The San Juans had not seen the likes of these three mines before, yet the big news was their neighbor, the Camp Bird, just over the range in the Imogene Basin of Ouray County. A record-smashing decade produced over $17,000,000, primarily gold, an amount never before or since equaled in the San Juans. Thomas Walsh was the magnate who brought it into millions. A shrewd, industrious Irish immigrant, Walsh already had made one fortune in mining and lost it in the crash of 1893. The claims that became the Camp Bird had been staked back in 1877, owned and operated by the old San Juan promoter William Weston, who sold them to a company which abandoned them as unprofitable in the 1880s. Weston and his contemporaries sought silver-lead veins; Walsh became convinced that the basin was rich in gold. He examined the area carefully and purchased property, starting a minor stampede by owners anxious to unload their claims. The last laugh proved his when a rich gold deposit was uncovered. Work started in the fall of 1896, and the mine was developed steadily until, by the turn of the century, Walsh owned a well-constructed modern plant, operated by electricity.

Walsh, the best-known San Juan mining millionaire of the twentieth century, developed his property conservatively. The money that poured into his bank account soon took the Walsh family out of the San Juans and into the eastern social circuit, to a home in Washington, D.C. As early as 1900, rumors cropped up that he was interested in selling the Camp Bird, which he finally did in 1902 for a reported $3.5 million in cash plus stocks and bonds. The noted mining engineer, John Hays Hammond, examined the property for the English purchasers and later managed it. Walsh told him that he had already made several million out of the mine and could no longer find anyone he could trust to manage it properly. Several lawsuits later (the plaintiffs claiming they had been cheated out of their claims), Walsh divested himself of the Camp Bird.

The Camp Bird Limited's operations reaped steady profits in the years that followed. The mine became the San Juans' most famous, even acquiring a European reputation. Jules Huret, a Frenchman touring the United States, could not think of going home without seeing it, although he was sorely disappointed at not finding gold hanging from the walls. His faith was restored somewhat when he saw a bullion bar at the mill. When labor trouble broke out all around him, Hammond kept the Camp Bird going. His own account of those hectic days stressed frank and fair dealings with his workers. Upon hearing that the union was coming to coerce his miners to strike, he called a meeting, discussed the issues, and was told by his men that they would solve the problem. According to Hammond, the confrontation between union and Camp Bird miners was short and to the point. "You damn butchers, what are you doing here? Go back where you came from. You're going over the cliff quick if you don't get the hell out of here!" For whatever reason, the Camp Bird stayed tranquil.

The mine and mill operation was a wonder of the region, savings in gold at the mill running well over ninety percent throughout these years. Stockholders certainly had no reason to complain; by December 1908, $4,600,000 in dividends had found their way into shareholders' pockets. The Camp Bird lived up to everyone's expectations, from London to Ouray. Almost single handedly it pushed Ouray County to the forefront of the San Juans by 1909, trailing only Teller and Lake counties in Colorado.[6]

With the Camp Bird, Tomboy, Smuggler-Union and Liberty Bell leading the way, the San Juans experienced their greatest decade ever, with gold production far out in front. After three frustrating decades of exploration and development, the San Juans claimed a place as one of Colorado's top mining regions (number two in 1909). Ouray, San Miguel, and San Juan counties headed the list. San Juan, which did not gain the same attention as the other two, produced steadily throughout these years without being dominated by any one of several large operations. The Guggenheims, of smelter fame, purchased Stoiber's Silver Lake property in 1901, the largest mining transaction to date in the county, roughly $2.3 million. The mine never did as well as they hoped. John Terry increased his Sunnyside operation until it was one of the county's best producers. Animas Forks, Eureka, Gladstone, and Chattanooga enjoyed moments of prosperity when a local mine or two hit a small bonanza.

A bird's eye view of the rest of the San Juans presents a rather dismal picture, with the exception of La Plata County. Though that county's production never equaled even a fair year for the Tomboy mine, it did top $500,000 in 1907, the best in La Plata's history. The

Neglected mine, west of Durango on Junction Creek, and the May Day, near the mouth of La Plata Canyon, spearheaded this little excitement. Mineral County's production was cut in half, to $1 million, Dolores averaged in the $100,000 range (maintained there by a large percentage of lead, copper, and zinc), and Hinsdale dropped to the same level, 1/6 of what it had been in 1899. Rio Grande (Summitville, basically) slipped nearly out of sight, much less than $10,000 per year. All of these districts were old and showed it. Small operations, working low-grade ore only seasonally, indicated the limited resources and did not augur well for a bright future, even for La Plata County.

In La Plata County, coal mining made steady strides. In the small company-controlled camps of Porter, Perins, and Hesperus and around Durango, the miners lived and worked. Production grew from 122,270 tons in 1900 to a decade peak of 189,357 tons in 1907, small compared to the major Colorado fields (state total 10,965,000 tons in 1907), but the largest the San Juans had yet mined. The trend was toward fewer companies, three in 1909 compared to seven in 1901, and more machine mining. Hand-operated small mines could not compete. The newest mineral in the San Juans was carnotite, yielding uranium and vanadium. Deposits were found from within a few miles of Telluride to the western fringe of San Miguel County. This rare metal at first had to be shipped outside to be worked, then a mill was built at Newmire (later Vanadium) below Telluride. The richest deposits were in the Paradox Valley, outside the San Juans as defined in this study. Its usefulness at this time was largely medical, with recovery of vanadium secondary. The market in the early 1900s was limited and mining proved spotty.

In both major and minor districts two trends came to dominate, consolidation and leasing. Obviously Telluride exemplified the former, as did Rico, where the United Rico Mining Company in 1901 consolidated many of the old producers into one operation. Leasing could spread the risk of loss among several groups, as Wells had done with the Smuggler-Union, or an entire mine could be leased, with a fee and royalties going to the owner. It was extremely infrequent that a mine in high-grade ore was leased; more typically the action was evidence of the owner's lack of finances, inability to operate, or loss of faith in his property.

Even in bonanza, San Juan mining confronted decades-old problems. The declining districts cried for investments and capitalists to ride to their rescue. Lawsuits crippled some of the mines for brief periods, poor surveying and lost claim markers causing most of the grief. Miners speculated that the ore became richer with depth and worried over mineral prices and smelting costs. Prospecting continued, though at a sharply reduced pace, and even turned up a few small mines that somehow had been overlooked during the past thirty years. This gave the local press something to crow about, before the discoveries pinched out.

Mining engineer R.M. Atwater, examining the Sunnyside for prospective buyers, professionally summarized the problems facing even the successful Terry. Freight rates were high, as "everything must be brought around the long way on narrow gauge lines." While the union situation was "very much improved over" the last year (1905) or so, "it is by no means safe or trustworthy." Mostly foreigners were employed in the mines, he concluded. Nor was the sixty percent recovery rate at the Sunnyside mill inviting; his clients would not buy.

When the American Smelting and Refining Company came on the field, Colorado smelting swung toward monopoly, a pattern reflected in the San Juans. This was strengthened further in 1901 when the Guggenheims gained control of the concern and brought in their smelters and organizational skill. Durango, Denver, and Pueblo all joined the smelter trust, leaving the rest at their mercy. Smelters still were being built, pyritic smelting (to treat low-grade iron-gold pyrites) being the newest innovation at the moment; Ouray and Silverton each secured one. Cyanidation gained more followers, particularly in the Telluride-Camp Bird area, and concentrating works retained their popularity. Only a few smelters failed to switch to cheaper and more dependable electricity.

In 1902 Thomas Rickard made some pertinent observations on San Juan smelting. He criticized the earlier efforts, because the smelters had been situated in localities ill-suited for

Near the Tomboy, the Smuggler-Union was only slightly less productive. High school boys often found employment as waiters or other above-ground work before going underground as miners. A mine this large had an excellent cooking crew, the management being well aware what made workers content.

The Gold Prince mill at Animas Forks looked impressive but failed because of poor planning, the wrong process, and not enough ore. Josephine Peirce remembered some great dances in the boarding house. "About ten men to each gal. Music was a fiddle and a melodeon. Square dances were the most popular, but waltz, two-step and polka were usual."

fuel and ore. They were, he felt, "doomed, therefore, to point a moral and adorn a melancholy tale." Even though these failures cost money and time, there had been some benefit. "In these early efforts there is a personal equation and a human interest lacking in the larger undertaking of later days because they represent the skill, hopefulness and energy of individual young men, many of whom have proved to be masters of the metallurgical art."[7]

While there were still those who experimented individually and left "melancholy" remains, smelting had become big business and was operated as such—the Guggenheims saw to that. The American Smelting and Refining plants offered ease of access, scientific management, and refined processes. The large mines usually had their own mills, which dealt with the smelters when necessary, and even smaller operations maintained concentrating plants. These factors placed the small, independent operators at a decided disadvantage, one which they could not easily overcome. Eventually they were forced out of business, a change with long-range implications for the San Juans and Colorado. Mine owners grumbled to themselves about the smelters and were not averse to complaining publicly in the press if they thought the problem warranted it. It usually proved to be an exercise in futility.

Grumbling about the weather proved equally as fruitless; with vengeance, several bad winters emphasized that times had not changed very much. A February snowslide in 1902 hit the Liberty Bell boarding house, killing sixteen men, including members of rescue parties caught in later slides. March was no better, and slides hit tramways, mills, homes, boarding houses, and other structures, killing several more people. The winters of 1906 and 1909 also were infamous, completely disrupting train service but fortunately causing fewer losses of life. The year 1909 produced twin disasters: snows and severe floods. The Telluride area was hit hard in September, when the town was cut off for several weeks.

Storms were part of the routine for the miner; fire was not. The worst fire in San Juan history hit the Smuggler-Union on November 20, 1901, when surface buildings at the Bullion Tunnel burned. Smoke and gas were sucked into the mine, killing twenty-four and overcoming others; tragically, doors constructed to prevent just that type of accident remained open. The stunned San Miguel Examiner, November 23, could only say that fortunately "very few married men" were among the victims. The financial loss to the company proved minimal, the loss of lives incalculable. The miner was now entirely a part of the corporation, his wages, his room and board, even the fate of his job determined by management. This loss of freedom underlay the laborer's uneasiness.

Thomas Walsh offered some good advice to owners and engineers in his 1908 School of Mines commencement address when he urged them to treat miners with humanity and justice and provide clean, comfortable quarters, wholesome food, and medicine for their use. The money would be "well spent" in gaining their appreciation and loyalty. "To use a mining phrase, you will be prospecting in human hearts, and may discover beauties of character little suspected." In regard to strikes he recommended a "heart to heart talk" with the men, fairly and squarely presenting management's case. By trying to reach the best side of their nature in dealing with them, he felt "strikes can nearly always be averted." Whether this latter advice proved useful to the freshly degreed mining engineers is questionable, but treating the men with humanity and respect certainly reaped benefits.

The complete corporation dominance after the turn of the century made it easier to find statistics and information on a miner's life and work. Wages, for example, standardized. In 1900 $3 per day was standard for miners in San Miguel County, $2.50 for outside laborers, and $4 for mill men (who worked longer hours); $1 was charged for board and lodging. Two years later the same wages were being paid to miners at Silverton and the Camp Bird for an eight-hour day, while laborers got $3 for nine hours and mill men $3.25 to $4.00 for twelve hours. West of Rico, at Dunton, miners received $2.50 and board. Specialization crept into their work. Charles Chase, superintendent of the Liberty Bell, wrote a friend that he needed a single-jack miner accustomed to drift work; no one else need apply. In 1902, for the entire

The Boston Coal and Fuel locomotive ran only a few miles out of Durango to the company's coal mine at Perins. In the right rear is the Strater Hotel, Durango's finest.

San Juans an estimated 7,301 men worked at mining, not including office help or those not directly involved with underground operations.

Coal miners were paid by the ton, 66¢ for each one mined. Mule drivers received $2.50 to $3 per day and boys who did various jobs got $1 to $1.50. Car loaders were paid $2 to $2.25 and outside laborers from $2 to $2.50. The coal mine workers generally received less for their work in an occupation with more danger than did their hardrock contemporaries.

In the all-important boarding house, cooks were paid by the number of men they served, starting at $2 per day for ten or fewer. Boarding house expenses for the larger mines were no small item; the Liberty Bell in 1901 spent $30,000 and still made a $9,000 profit. Stockholders appreciated that; miners were not so sure about the justice involved. Though complaints were heard, the food overall must have been at least edible. Rickard called it "surprisingly good as a rule," the weak spot being coffee, a concoction better "adapted for staining floors or removing boiler scale." Another observer described the dinner hour. When everything was ready, the doors swung open, men rushed in and attacked the food with no restraint or manners, "ravenous the only word for it." Care was also taken with the miners' bunk houses, the best being roomy, electrically lit, steam heated, and furnished with bathtubs and running water. The Smuggler-Union boarding house featured a large lounging room with an open fireplace, all "cozy and homelike."

Men working in wet places, "where rubber coats" were necessary, received an extra 25¢ per shift. All miners went to and from work (portal to work site) on their own time. Realizing the need for avoiding sickness, some of the larger companies built bathing and drying rooms at the portal to help the men avoid colds and the dreaded pneumonia. Despite these precautions, a deadly killer roamed the mines—silicosis, or as they called it, "miners' consumption." An alarming increase in deaths related to it appeared after 1900, partly because it was medically recognized, but mainly because of the deadliness of the early machine drills, which lacked a flow of water to drown the rapier-sharp dust ground out of the rock. Miners also fell victim to cave-ins, premature blasts, missed holes, electrocution (coming into contact with a live wire), falling down shafts, and their own carelessness. Some seemed to resent the new machines, and others lacked the technical skill to utilize them effectively. A new piece of machinery in the hands of a miner with a prejudice against it was soon abandoned. Gradually he came to realize the labor and time-saving capabilities; once the drills became safer, acceptance followed. Labor-saving devices could only go so far, however; even in the best operated and equipped mines, "men worked hard, often under trying and dangerous conditions."[8]

After a couple months of intensive labor, the miner had to get out and "blow wind," as one who worked in the mines explained it. The mines did not generally shut down now for Christmas holidays; Labor Day in early September was the miners' holiday. The unions strongly backed it, and it caught on as a family celebration, with games, speeches, dances, and other activities. The miners "blew wind" for their own day. Down in the camps waited others ready to help them "blow wind" anytime, as long as their money lasted.

[1] James Smith, *8th Biennial Report of the Bureau of Labor Statistics* (Denver: Smith-Brooks Printing Company, 1902,) page 278. *Criminal Record of the Western Federation of Miners: Coeur D'Alene to Cripple Creek* (Colorado Springs: Colorado Mine Operator's Association, 1904), page 31. *The Miners' Magazine*, August 1903, page 4. See also *A Report of Labor Disturbances in the State of Colorado from 1880-1904 inclusive.*

[2] For general information on the troubles see Smith, *8th Biennial Report;* Mark Wyman, *Hard Rock Epic* (Berkeley: University of California 1979). *Criminal Record; A Report of Labor Disturbances*; Roger N. Williams, *The Great Telluride Strike* (Telluride: NP, 1977); and George Suggs, *Colorado's War on Militant Unionism* (Detroit: Wayne State, 1972). The *Miners' Magazine*, 1901 and 1903-04. *San Miguel Examiner*, 1901-04. *Daily Journal*, 1902-03. *Engineering and Mining Journal*, 1901-04. Sherman Bell, *Biennial Report of the Adjutant General* (Denver: Smith-Brooks, 1904). Gene Gressley (ed.), *Bostonians and Bullion, the Journal of Robert Livermore 1892-1915* (Lincoln: University of Nebraska Press, 1968), pages 97-99. Martin Wenger Manuscript, Center of Southwest Studies. Bulkeley Wells to William Bell, September 14, 1903, Bell Papers, Colorado Historical Society.

[3] The *Miners' Magazine*, December 3, 1903, page 9. See also 1904 especially. Ray S. Baker, "The Reign of Lawlessness," *McClure's Magazine*, May 1904, pages 43-57. Bell, *Biennial Report*, pages 23 and 202-217. *Report of the Smuggler-Union Mining Company for 1904* (Boston: NP, 1904), pages 5-9. Robert Livermore Speech, April 1, 1935, Livermore Collection, University of Wyoming. Suggs, *Colorado's War*, pages 144-145.

[4] The Archives of the Western Federation of Miners, Western History Collections, University of Colorado, contain records from these locals. The *Miners' Magazine*, 1901-04. *Engineering and Mining Journal*, 1901-09.

[5] Botkin to author, November 7 and 29, 1972. For the Tomboy, Smuggler, and Liberty Bell see: *Engineering and Mining Journal*, 1900-09; *San Miguel Examiner*, 1900-09; Tomboy records, Hague Collection; F. Bradley, *The Tomboy Gold Mines Company, Limited* (London: Growther and Goodman, 1906); *Report of the Smuggler-Union Company* (1904 and 1905); E.B. Adams, *My Association with a Glamorous Man . . . Bulkeley Wells* (1961); Gressley, *Bostonians*, page 153; Henderson, *Mining in Colorado*, pages 224-225.

[6] Camp Bird material is plentiful. For Walsh's account see Thomas Walsh, "Commencement Address," *Quarterly of the Colorado School of Mines* (July 1908), pages 13-15. John Hays Hammond, *Autobiography* (New York: Farrar & Rinehard, 1935), volume 2, pages 483-494. Jules Huret, *De San Francisco Au Canada* (Paris: Bibliotheque-Charpentier, 1917), pages 188-205. R.J. Frecheville, *Report on the Camp Bird Mines* (Ouray: NP, 1904). Rickard, *Retrospect*, pages 76-77. *Engineering and Mining Journal*, 1897-1909. *Ouray Herald* and the *Plaindealer*, various dates 1900-09.

[7] T.A. Rickard, *Across the San Juan Mountains* (New York: *Engineering and Mining Journal*, 1903), pages 56-60. R.M. Atwater, "Report on the Sunnyside," copy loaned the author by Allan Bird.

[8] Harriet Backus to author, June 27, 1973. Ernest Hoffman, Interview, March 22, 1973. Thomas Walsh, "Commencement Address," page 16. Harry Lee, *Report of the State Bureau of Mines for the Years 1901-02* (Denver: Smith Brooks, 1903), pages 245-246. Chase to Fred Carroll, March 26, 1904, Atlas Mining and Milling Company Papers, Western Historical Collections. *Engineering and Mining Journal*, 1900-09, has much on working conditions and wages; so do the local papers. Report on Emma Gold Mining Company, Hague Collection, Huntington Library. Silver Lake Pay Book, Center of Southwest Studies. *Liberty Bell* (annual report 1901), page 6. Edward Pierce, "Telluride in 1900," *Pioneers of the San Juan*, volume III, page 30. Rickard, *Across the San Juan*, page 13.

"A Harvest Gathered Once"

The San Juans were never noted for placers; the photograph illustrates some of the problems encountered in sluicing. These men were working in John Moss's ill-fated La Plata Canyon endeavor.

Mining gets into one's blood—the adventure, the fascination, the expectation, perhaps even the danger. In spite of the hardships and the risks, people are drawn to mining as moths to the candle's flame. Nicholas Creede, interviewed a short while after his famous discovery, commented: "I never cared a great deal for money, but had a great desire to find a great mine—something that would excite the world . . ." He went on to say, "there is a great fascination about the life of a prospector for me". Other less fortunate San Juaners echoed his sentiments. As a result, they prospected, dug, blasted, burrowed, and carted away tons of ore, all in the name of mining.

Some critics argued that mining was not a legitimate industry or business. Thomas Rickard sharply disagreed: "There is as much luck in mining as in any other business enterprise, hardly more; there is as much room for skill and sense in mining as in other commercial undertakings and a good deal more."

The photographs that follow feature the men, the equipment, and the mines that are gone now. Rickard provides an epitaph: "Mines are shortlived. They yield a harvest that is gathered once only." From their depths came gold, silver, and other metals that have enriched a region, state, nation, and even places beyond. They are gone but hardly forgotten. A writer described a mine in the April 1, 1882 issue of the *Mining Record* as

> . . . a living being. The shafts and winzes are its lung through which it breathes. The pipes or tubes of its pumps constitute a system of circulation. It lives upon the fuel to its engines, and it has a name, a personality and even a sex, for with the miner as with the sailor in regard to his ship, the mine is *she*. From her are born products which enrich mankind.

(this page) "Ho, for the San Juans" was never easy over trails like this one. William Henry Jackson's photograph illustrates why it was costly to freight supplies and equipment. Fortunately, time was less a factor then than it is now; a season might pass before a mill could be laboriously moved over such trails.

(opp. page, top) Jackson also photographed these pioneering miners above Cunningham Gulch in 1875. The equipment, attire, and determination were typical of the day.

(opp. page, bottom) These men mined coal and lived in a company town at Perins, directly west of Durango. Most of them probably came from Eastern Europe. Their work was more dangerous and less well paid, but without coal the San Juan fuel situation would have been serious.

(top) Dave Swickhimer (third from left, top row) and his Enterprise crew. They proudly surround ore sacked and ready for shipment, part of the 2,000 tons that piled up in the spring of 1890 when weather made the roads impassable. Several of the miners are wearing slickers, indicating the wetness of this mine.

(bottom) Underground work in any mine was dirty, noisy, and difficult. A skillfully laid track had a slight grade so that full cars were pushed downhill and empty ones up. This was the Sunnyside mine, No. 2 crosscut, above Eureka.

(top) The compressed air-driven drill reduced the physical labor and time involved in drilling but produced an increased hazard of rock dust. The early models were called "widow makers"; the introduction of water cut the dust danger. This crew in the Gold King Extension near Gladstone operated a Rand drill. Note the primitive hard hats and the use of candles.

(bottom) The Wheel of Fortune crew in 1877, for the time a fairly large one. Located in the Mt. Sneffels district south of Ouray, this mine once sold for a reported $150,000 and was considered one of the finest ever opened in the state."

SILVER LAKE MINE.

(top) From the Wheel of Fortune to the Silver Lake mine crew spans the time from the days of individualism to corporation control. Everything in this picture, from buildings to men, indicates the change which had come about. The cooks look to be a determined lot; so do the miners.

(bottom) Mining in the high reaches of the San Juans was physically hard and isolated. In September 1886, Marshall Basin was a decade away from the greatness it achieved when the Smuggler-Union, a consolidation of several properties, came to the forefront. The buildings are located at over 12,000 feet, far above the trees that provide shelter from the winter winds.

No miner could work underground without the support of a surface crew, which ranged from the common laborer to the engineer running the hoist. This is the machine shop of the Camp Bird after 1900; it was equipped far better than most San Juan mines.

(this page) Winter freighting, always dangerous, less dependable, and more demanding, was a major incentive for the introduction of the year-round tram. This scene could have taken place any year, but it is about 1914 outside of Ouray, showing that the method had not changed much in fifty years.

(opp. page, top) The Ute-Ulay was Lake City's best mine; the little camp of Henson grew up around it. Its production, however, never equaled the hopes of its owners. This late 1870 or early 1880 photograph shows a typical larger mine layout for the period; nearby trees had already disappeared in the scramble for wood.

(opp. page, bottom) One of the factors which cut the cost of mining was the introduction of electricity. The first AC generating plant was built at Ames a few years after this photograph. Ames was typical of the small San Juan mining camps; it basked for a brief moment in the sun, then rapidly declined as local mines failed to match expectations. The patient burro proved to be an unsung hero of the San Juans.

(top) The ultimate in transportation was the railroad, or so thought the nineteenth-century San Juaner. Tremendous engineering problems hindered construction. Otto Mears built the Silverton Railroad into Red Mountain and beyond. A wye had to be constructed around the depot, seen in the center of the photograph, to overcome site limitations. The National Belle mine is in the upper background.

(bottom) Tramways were one of the engineering marvels of the San Juans; they went almost anywhere, over or around obstacles. Without them the cost of mining would have been prohibitive, especially the expense of freighting. This crew is building a tower for Creede's Amethyst mine.

(top) Neither the Black Wonder mill nor the camp of Sherman lasted for long. Almost every district and many mines thought they needed a smelter or mill. A fatal waste of money ensued before the concept of regional smelters took hold. One observer called them "moments of ignorance and folly."

(bottom) If the trails were good enough, the freight wagon replaced the burro. Wagons lumbered from mine to mill or train siding and carried back provisions and occasional passengers. San Juan trails tested the drivers' skill and courage. This group stopped near Creede.

(top) Logging was as essential to the success of mining as smelting. Without the logger cutting and hauling the timbers and the sawmill turning out the lumber, mining and settlement could not have succeeded. This log was hauled to the American Nettie mine outside Ouray, a mine notorious for engendering silicosis among its miners.

(bottom) The Gold King mill was a large operation after the turn of the century; even the Silverton, Gladstone, and Northerly Railroad ran to its doorstep. Several earlier mills had been on or near this site. Gladstone buildings appear in the lower part of the photograph.

(top) Individual opportunity disappeared, leaving the miner to earn a daily wage. Suddenly mining became less appealing and the pay seemed to be unfair. Absentee ownership took away the camaraderie of an earlier day and tensions mounted. As a result, the Tomboy hired night and day guards to protect its property. Bob Meldrum sits on his horse in the middle. The fellow reading the paper is a nephew of one of the English owners; he received a prize for having read that English paper in the most remote place and greatest distance from England.

(bottom) Ominously, both miners and owners hardened their attitudes. Finally violence broke out, centering in the Telluride district. Troops moved in and garrisoned the region; management eventually won. The price was heavy; never again would feelings be the same and the last vestiges of an earlier time disappeared. These guardsmen were billeted at the Smuggler-Union boarding house.

No man made more money in San
Juan mining than Thomas Walsh, shown
here at his desk in the Camp Bird office.
By the turn of the century, the cost and
large scale operations had ended the era of
individual opportunity. A few fortunate
men like Walsh made money, but most
profits went into corporation coffers.

The term "twentieth" century sounded strange to the ears of those who had grown up in the nineteenth. Opportunities beckoned and new challenges arose for young people; for the middle-aged and older, it was as if they had aged in a day, and suddenly found the world of their youth slipping beyond reach. The editor of the *Lake City Times* did not bother to compose an editorial himself as 1899 passed into history. He apparently felt that the one he took from the Colorado Springs *Telegraph* sufficed:

Hail the New, Farewell the Old

One can go on asking questions about the year 2000 all day long, but for us the dawn of another hundred years opens up such a magnificent vista of possibilities and opportunities so far ahead of those which confronted the people who saw the dawn of 1800 that we may well devote our attention to the work that lies before us and let the future generations work out their own salvation.

Young Lake Citian Frank Hough disagreed with his newspaper; he thought the century ended on midnight December 31, 1900, at which time he recorded in his diary there was a "whole lot of noise with whistles and bells and shooting."

The year chosen to mark the turn of the century mattered little, except that it put the pioneering period behind as surely as the number nineteen replaced eighteen on the calendar. The San Juaner of the twentieth century was a generation removed from the 1870s—so far removed that when an "old timer" returned to town or died it was news-worthy. The Pioneers of the San Juan continued to hold yearly reunions, amid thinning ranks, as did those even older veterans of the Civil War. Occasionally, editors felt obligated to remind readers that Decoration Day was approaching, a day set aside to honor soldiers and memories of that war, "it does not matter upon which side they fought." August 1, Colorado Day, prompted editorial pens to eulogize the state's founding fathers and slip in a bit of Colorado pride and propaganda. The original San Juaners stood in the ranks of those founding fathers; some had even mustered with the legendary boys of blue and gray. In 1901 Ben Bascom, a pioneer Lake City businessman, returned to his old stomping ground for the first time in over a decade. Bascom grasped the changes when he observed that while the town had not changed much, "many once familiar faces" were missing, and "many, many new ones" were present.[1]

Man's allotted three score and ten have passed since then, and all Bascom's once familiar faces, as well as most of the new faces, have departed. Youth has long since abandoned those remaining. As they reminisced about those years, these old San Juaners have allowed us a delightfully personal glimpse of an age that is hauntingly familiar, yet different from our own.[2]

Much of what they recall centers on growing up and school days. Two Telluride sisters spent many happy hours making bouquets of clover blossoms in their backyard, taking time out for a carefully balanced walk on top of the board fence around their house. At school they played hopscotch, "both round and straight." A Creede lady remembered her frame school building and that she had to go away to high school because Creede had none. Norman Bawden, a Silverton newsboy, made a small fortune when snowslides barricaded the railroad;

he and a friend dragged a toboggan into town heaped with papers, which sold for as much as one dollar per copy to the news-starved townspeople. Drilling contests also stand out in his memory. Before the excitement of the contest, blacksmiths carefully hardened and sharpened the drills. He once saw a "world champion drilling team," or so it was billed. "Maybe they stretched the truth a bit, but they would have been mighty hard to beat."

Ernie Hoffman also grew up in Silverton. His first remembered impression was of many people on a busy main street, a fascinating scene to his innocent eyes, but a problem to parents frantically searching for a lost son. School was "pretty strict," no permissiveness; if Ernie got into trouble at school, he automatically answered for it at home. There was "no question about that." Plenty of jobs were available for boys, especially in the summer, when outside work was available at the mines (never underground) and in the kitchens of the boarding houses.

Carrie Dyer taught school in Silverton in 1909 and remembered the times much as Hoffman did. The people were friendly, and nearly every day a pan of hot rolls or a pie from some pupil's parents would appear at the door of the house she and three other teachers rented. The teachers were included in the "home-made amusements," even skiing in long dresses. With money abundant and the mines working, Silverton was a lively and interesting place that year for the young teacher.

"Every damn nationality" crowded into town, commented Ernie Hoffman: Finns, Swedes, Norwegians, Cousin Jacks, Austrians, Italians. "We considered them all Americans—just like the rest of us." There was a Cousin Jacks' saloon and a Swedish saloon, and they all had their dances, "those Swedes were hell on the big feeds." The 1910 census confirms Hoffman's observation for Silverton and for the entire San Juans, the percentages of foreign born running from twelve to forty-three. In San Juan County (seventy percent of the population was in Silverton), foreign-born whites accounted for forty-three percent of the total in 1910, up three percent from 1900. A breakdown found Italians and Austrians way out in front, trailed by Swedes and English. In neighboring San Miguel, a third of the population was foreign born. Here the Finns, Austrians, Italians, and Swedes predominated. The other of the big three, Ouray, showed a quarter foreign born, with Austrians, Swedes, Italians, and English leading. A lesser district, Hinsdale County, had only eighteen percent foreign born, with Swedes and Canadians heading a numerically small foreign contingent.[3] A rule of thumb: the more prosperous the mines, the more foreigners were to be found. Obviously the eastern European immigrant wave had reached the San Juans, to a great measure centering in the three major districts.

The impact of these new immigrants was evident in Hoffman's remark about ethnic saloons, and in parties and unique customs (an Italian christening, for instance), growth of the Catholic church (a concern of the Protestants), and small ethnic clusters within the communities. They did not break the social or political barrier immediately. The Chinese still faced discrimination, Silverton, Telluride, and Ouray forcing them out in 1901-1902 by boycotting their businesses. The unions partly promoted this in their desire for American laborers only; some businessmen likewise encouraged their removal to eliminate competition. Brief international tension flared when the Chinese minister in Washington protested to the American Secretary of State. He, in turn, notified the governor, who squelched the matter by reporting there was nothing that could be done, because nothing illegal had been perpetrated on the Chinese. In May 1902, however, a Chinese restaurant and laundry were looted and the owners driven out of Silverton. Passions eventually cooled and by mid-decade the Orientals were back in Ouray and Telluride. Ouray even came to the support of Wing Kee when he was threatened with deportation by the United States government.

Frank Hough took little notice of the changing nationalities around him. His Lake City diaries reflect a middle-class routine. His life in 1900-1901 centered on his family, social activities, and the church. An exciting day for him brought his new bicycle, "a bird, don't you forget it." He went to church business meetings and prayer services, played dominoes with friends, made ice cream, shivareed a young married couple, practiced with the choir, and attended dances. "Today was a merry day," he wrote in 1900. "We ought to rejoice of having such a nice Christmas." The next day at work he felt pretty "rocky," having been up until two o'clock in the morning. July Fourth started early in Lake City, with plenty of fire crackers. In 1900 a picnic and baseball enlivened the afternoon, and a ball and supper the evening. Frank enjoyed bicycling, hikes, and croquet during the summer, and dances year round. His mother did not approve of his social whirl, Frank confided after a Leap Year ball: "Mother was kind of hot or mad because I went to dance last night and wouldn't hardly speak to me all day. That is alright am my own boss."

Unlike most of his neighbors, Hough was a Republican, an active one who helped organize a McKinley and Roosevelt club. He went to Denver in September 1900 to attend the state Republican convention, using his spare time to see several plays, ride a trolley, and listen to Senator Henry Cabot Lodge speak. The campaign picked up, "red hot on both sides." On election day he served as a judge. William Jennings Bryan did well locally, as expected; two more days passed before the Denver papers arrived with the complete national returns. On November 8, Hough gleefully noted, "when they saw them this evening they were disgusted to see that Bryan was defeated and McKinley was elected."

Compared to the 1890s, politics had lost some of their fervor with the demise of the Populists and the defusing of the silver issue. Party lines gradually returned to the more traditional Republican versus Democratic squabbles over local issues. The latter party fared much the better; Republican silver heresy was not easily forgotten. Telluride, showing the unsettled, bitter effects of the strike, gave traditionalists like Hough a scare. The Socialists fielded a ticket there and also at Ophir for several years. They elected a few local officials but generally were defeated by a nonpartisan ticket. The Socialist candidate for governor tallied 297 votes at Telluride during the height of the tension in 1902. As proven in the 1890s, San Juaners could be radical, primarily in defense of their personal conservative interests. No unified front appeared after 1896, however.

If politics caused a temporary row, the intriguing red-light district raised a continual one. Visitors and residents, young and old, were attracted by it, while reformers were repelled. Ernie Hoffman remembered that the district was off limits to youngsters; after only one warning from the "cop" about hanging around, the culprit was personally escorted home. Saloons were also off limits, but minors could go to the local beer garden and brewery, where there was soda pop of all kinds and, if they were lucky, a sample of the "dutch treat lunch." Over in "picturesque" Telluride, as recent college graduate Robert Livermore described it, the mine operators believed the red-light district to be a benefit. Livermore was convinced that was because the "sooner the miners got broke," the quicker they returned to work. The Christmas of 1903 stood out in Livermore's memory, with every saloon (forty or so) offering a steaming bowl of Tom and Jerry free to all and the "line" decorated with mistletoe and lights.

The age-old controversy raged on about the merits of the line, neither side conceding an inch. From the vantage point of sixty-plus years, Hoffman thought that "the girls" had been good for Silverton; they were in it for money, "just like anybody else." The fact that the situation was "wide open" reduced sex crimes, and the town required weekly examinations to prevent venereal disease (Lake City demanded semimonthly exams at the prostitutes' expense). Livermore's wife hired a graduate of the "row" as a maid, who proved to be an excellent cook of exemplary character; other local wives were not so liberal in their attitudes.

According to the opposition, prostitution was sin with a capital S and must be rooted out. Hindsight makes it easier to see now, than at the time, that the battle was being won by this side. Back in the nineties, the tide had changed; the "women of the town," the dance halls, and finally even the saloons were gradually being segregated. The next step was complete

Rico was a declining community by the time this photograph was taken (approximately 1906). Only one more brief mining revival was yet to come. The Chas. Engel Mercantile Company carried a little bit of everything, typical of a mining community store.

Even after the turn of the century, with automobiles and trucks just around the corner, horses and mules moved the freight. Sometimes they pushed and pulled to get heavy cable to the mines.

prohibition of this ancient profession and the other evils.

On a different front, another battle was being won. Crime abated noticeably in all except the largest towns. Rico's police magistrate's docket in 1902 listed mostly just monthly fines against prostitutes and gambling, strongly suggesting that their illegal existence was winked at for the sake of city revenue (Lake City appears guilty of the same policy). Silverton was obviously guilty, regularly fining the girls and gamblers on the first of every month. By 1907-1908 this practice of fining had ended in Rico, and the docket listed fighting, disturbance of the peace, fast riding (Telluride limited the speed of bicycles and "riding machines" to 8 mph and horses to 10), as well as one prostitute and two gambling violations.[4]

On the subject of uncondoned behavior, a married Silverton woman, identified only as Madelin, was having an affair with Anthony Linquist, a miner who lived outside of town. Her letters expressed deep feelings: "We will always be one, until the time comes when we can be free about it," and concern about his occupation, "Take the best care of yourself so nothing happens to you—don't get in no *slide* or don't *slip* and *fall*, go *slow* and take your time, for you have something to live for." Ever discreet, she admonished her lover, "Dear never keep these letters—as soon as you read them burn them up for we don't want *no one* know our business—for it is best—unknowing facts is the winner."

One-time Silverton banker Guy Emerson related his memories of those years. Life was interesting and more respectable for him:

> In the summer business men fished in Mineral Creek, Molas Lake and elsewhere, and in the winter skiing was good. I was a comparatively young man in those days, and for entertainment on Sundays, I would take the D&RG narrow gauge train leaving Silverton in the morning, down to Tacoma; off there go up in the hill on the old tramway; walk at least 3 miles on the way to Lake Electra; fish all day; lots of good trout; walk back down to the brow on the hill; take the old tram down to the station at Tacoma; and wait for an indefinite time for the passenger train from Durango; get to Silverton sometime during the night; and think I had a good time.

In the twilight of the Victorian years, life was full of contrasts.

During this decade, Americans were titillated, shocked, and horrified by the revelations of the muckrakers about the life they had considered so pristine. Typically optimistic, they launched the Progressive era, a multifaceted effort toward clean-up and reform. The San Juans evidenced little of this movement, except in the moral sphere, where progressivism dovetailed with several local issues. Cigarettes drew the wrath of local reformers. Lake City decreed that no boys under sixteen (nice girls would not think of it) could purchase tobacco or cigarettes, and Telluride's school board and teachers fought a dogged, though losing, battle against the smoking "contagion." The *San Miguel Examiner*, November 6, 1909, was moved to editorialize that the smoking habit "could and should be overcome, not only in children, but men as well."

Telluride compiled, updated, and published its city ordinances in 1909. Evidence of a change in public attitudes was ordinances specifically prohibiting such unacceptable, indecent practices as appearing in a state of nudity, dressing as a member of the opposite sex, and selling lewd and obscene books. Prostitution, open adultery, enticing a minor into a saloon, beer hall, or place of "immoral or indecent tendencies," or selling liquor to women were all banned. It was the "fallen" women who would drink in public they were after. Fraudulent weights and cruelty to animals likewise were offenses, as were "opium joints." Thus did the city recognize the existence of one of the gray areas of mining communities, drugs.

The *Ouray Herald* explicitly described both the changing times and reform sentiments, when it remarked on the council's closing of dance halls:

> [This] marks the beginning of a new and more progressive era in the social conditions of Ouray . . . . Dance halls are the product of new mining camps in the mountains or boom towns on the border of civilization, and Ouray has long since outlived the excuse for such institutions, if indeed, there ever was any.

George Puth Collection, Fort Lewis College

(right) Unlike Rico, Eureka's future was still ahead of it. Terry's Mill is seen in the background and a Silverton Northern train at the right center. Most of the open space would be filled in by the time of the World War I boom.

(below) Once the unions began their rise, Labor Day became a major holiday. It included baseball games, a parade, various contests, a dance—the possibilities were endless. Silverton turned out well for its parade, and the float was typical of the homemade variety of this era.

First Federal Savings

Reasons for this action, after all these years, were several: dance halls had become the "frequent resort" of young men, many in their teens; management had overridden police regulation; and the establishments were "an eyesore and gave a black eye to Ouray." Tourists arriving on the train had to pass by them on their way into the city, receiving an unfortunate first impression. The proprietors promptly scurried to reopen in Telluride, where they encountered a cool reception by the press. The closing of the Ouray dance halls dumped a lot of "superannuated fairies and nondescript musicians" into town, moaned a reporter.

What could happen to the best intentioned crusades was exemplified in Telluride. In 1901-1902 all the local dance halls had been closed, because of "unlawful pursuits and practices." Within a month a citizens' petition to reopen forced the council to hold a special December meeting, but they held firmly to their decision. Following a dance-hall-less holiday season, the council took under advisement another petition, finally bowing to public pressure in February and revoking its earlier resolution. The councilmen soothed their consciences by decreeing that the dance hall proprietors must pay the salary of an extra marshal, who would devote most of his time to policing those dance halls. In 1907 Telluride and Creede, in a fit of reform, closed the gambling games for a while. Ouray had done the same four years before, removing the slot machines; a plot of the WCTU, wailed an objector. WCTU pressure also promoted renewed interest in stopping the sale of liquor to "habitual drunkards."

The Women's Christian Temperance Union—there was a militant enemy of the old ways. Its primary target was that "great ailment," liquor, but it went further, promising that prohibition would also mean the end of "questionable districts." Its members still hoped to convince their neighbors of the rightness of their cause without resorting to political force: "It is not revolution we need or want, but Divine evolution." The miners greeted the whole idea with something less than enthusiasm. San Juaners would toe the mark only when statewide prohibition came some time later. The churches allied themselves with the WCTU in this fight; indeed, their membership rolls duplicated each other. As a first step in this direction, churches favored a Sunday closing law, which became more common on the statute books than was its enforcement on the streets. Blanket closing of all stores proved exceedingly controversial, the closing of saloons less so—they felt the wrath earliest and hardest.[5] The church, meanwhile, found its role more and more circumscribed by materialism, growing maturity of the community, and, in some cases, by the camp's and district's decline. From a community force it receded to a congregational force, with limited impact beyond its walls. This was to be expected and did not detract from what the church accomplished at the time.

The WCTU was not the only area in which women were active; they continued to make strides in the business world and even broke the sports barrier. The St. Louis Bloomer Girl baseball team lost 6-5 to the Telluride Elks team in 1908. After the starting pitcher was battered for four runs in the first inning, a male replaced her. In the San Juans, as elsewhere, women faced discrimination; less pay for equal work was a heavy handicap. They still had a long way to go to achieve equality, even with the vote and new power.

Tucked safely away from militant virtues and vices, at least until they ventured to town, were those people who lived at the mines. Harriet Backus has charmingly described her life as a young wife in Tomboy Bride. She and her friends in Savage Basin, high above Telluride, inhabited their own world, connected only by wire and a curving road to the outside. In her years there, 1906 to 1910, about twelve couples and 275 to 300 miners were in residence; the cabins were situated "quite a ways" from the boarding house. Although she did not get to know the miners, and only second-handedly their work, she always found them courteous. It did not take Harriet long to learn that a miner's language was "not what I was used to and I had sense enough to ignore it." Learning to cook at that elevation proved challenging, moments of despair offset by highly successful ones. When a woman became pregnant, it was customary to hide her condition. Mrs. Backus did that by wearing a very full dress, which fell straight from shoulders to floor. The Tomboy was no place to have a baby, and before it was due she traveled to Telluride and stayed in a rented room until time for her con-finement. She missed a good opportunity to observe the town, because custom also decreed that she stay hidden in order "that her condition would not be noticed." A month after her daughter was born in the Telluride hospital, Harriet returned up the long trail to home, where the company doctor was much more experienced with miners' injuries than with pediatrics.

Home remedies and patent medicines still took care of most of the San Juaners' aches and pains. A doctor was simply too far away, or the ailment did not seem serious enough to warrant his services. In this day before regulation, newspapers carried advertisements for miracle medicines for only a few cents. Lydia Pinkham's vegetable compound took care of all "those peculiar ailments of women;" after all, a "woman best understands a woman's ills." Then there were those like "Dr. King's New Discovery," absolutely guaranteed to cure coughs, colds, la grippe, and all throat and lung troubles. From the blatancy of the appeals, it appeared that San Juaners suffered from catarrh, headaches, nervousness, rheumatism, sore tired feet, and a host of skin troubles. They also fell victim to listlessness, diarrhea, psoriasis, and constipation. For men who came down with "social problems," or feared they were los-ing their "manliness," doctors promised a cure through the mail. Whether they believed all they read, few San Juaners were without their favorite remedy.

For social companionship the Backuses turned to the other families whose breadwinners also worked on the management side of the company operation, stories of whom she delight-fully recounts in her book. There was no church, so Sunday school was first held in the school building, then in the YMCA—the highest "Y" in the world.

The YMCA had come, as Mrs. Backus remarked, to "bring forth a little religion" to the miners and improve their social and cultural opportunities. The Smuggler-Union and Tom-boy both had branches, with bowling alleys, pool tables, and shuffleboards. The Tomboy was the larger of the two, having 127 members in 1908. The Smuggler branch closed a year later and was replaced by a company-sponsored organization on more "liberal lines" (unex-plained). The miners at these larger mines led a far different life, far more removed than simply in years, from their earlier brethren.

When the "Y" entered, it boasted, "booze usually takes to the tall uncut." Liquor proved to be no problem at the Tomboy, as Harriet Backus pointed out, since none was allowed, except for special occasions. For the miner who wished to be "in his cups" or to see the girls, it meant a trip to Telluride after the paycheck was drawn. Livermore's observation that the owners believed the red-light district to be a benefit, because broke miners quickly returned to work, probably did not miss the mark by far.

Politics did not arouse much passion among the Tomboy people; in fact, Mrs. Backus laughed when asked how they voted. She did not think anyone there gave voting much con-sideration; being so far from political groups, "we did not have the interest." The *San Miguel Examiner* concurred, blaming the decreased vote total in the 1907 city election on the many miners who did not leave work to vote. "Several ladies did not think it worthwhile to exercise the franchise either," chided the male reporter.

The Tomboy school, only one large room, convened for a four- or five-month session, according to Harriet. To Mildred Ekman it was the center of her young life. During one win-ter, her folks lived in a house near the powder magazine, which was always separated from the main workings, for obvious reasons. She and her friends would hitch a ride with the powder monkey in the morning, riding in style on top of the powder boxes. When there were parties, no one worried about babysitting for the children. Sleepy tykes were piled on top of coats until everyone was ready to go home, then parents claimed their own. For entertain-ment, throwing rocks down old shafts and listening to them bounce served to pass many hours, as did rock collecting. This latter hobby of Mildred's caused her father to be accused of highgrading, when a suspicious pile of rocks accumulated around his home. Fortunately, a quick investigation by the manager determined the culprit. Mrs. Ekman also remembered when her baby sister died, and her parents took the body down to Telluride through a snowstorm.

Alex Botkin, a friend and co-worker of the Backuses, was one of the best athletes at the Tomboy and the undisputed champion of the local tennis court. Playing at 11,500 feet elevation was no mean feat. Botkin also contributed his talents to the company baseball team, which had no practice field in Savage Basin. One well-remembered diamond possessed a fairly level infield, but had tree stumps as hazards for the outfielders. On a windy June day in 1908, the Tomboys went into Telluride and dropped a 14-16 game, Botkin playing shortstop. The Liberty Bell and Smuggler-Union also had nines, the former actually fielding two teams, from both mine and mill.

The general store at the Tomboy carried such clothing as the miners needed, tobacco, and various patent remedies. The proprietor also collected the mail and picked up the incoming deliveries. Botkin remembered the store as a popular spot for gambling, a favorite amusement even when the YMCA was going strong, and one that the miners would not give up. Looking back over their years there, both Backus and Botkin agreed that they enjoyed the life, Botkin considering it "very interesting," with occasional trying moments. One of those moments occurred when the company staff held a party for some Telluride friends. Unfortunately, the shift bosses had not been invited, and the next morning Botkin faced an irate, elaborately dressed wife who stormed, "I want to know why the shift bosses and their wives were not invited to the dance." Botkin explained that they wanted to have a balance between hosts and guests, whereupon his antagonist rejoined, "Well, Mr. Botkin, let me tell you that where Henry and me come from we were used to going with the best," and marched out.

James Batcheller likewise worked at the Tomboy during these years, as surveyor and mill superintendent. Writing his friends, the Backuses, years afterward, Batcheller said both he and his wife were grateful for having had the privilege of "living such an unusually active, eventful and interesting life before settling down." Another picture stayed in his mind—it had sickened him to watch some of his slovenly neighbors heave garbage out the door all winter long, where it piled up to await the spring thaw, warm days, and flies. Remembering the majestic mountains, he concluded, "Where every prospect pleases, and only man is vile!"[6]

"Only man is vile"—not all the memories and recollections of these early 1900s dwelled on the good and pleasant. Margaret Miller reviewed her youth in Creede and minced no words:

By the time I was old enough to have lasting impressions Creede was only a small town. As I think back, what comes to mind most clearly is the tragedy in the place. I remember:

Women weeping because there was a bad accident in a mine. Strong men suffering a slow death from 'miners' consumption. A girl on Sixth Street who took too many pills because her life was unbearable [Sixth Street was the red-light district]. Children poorly clothed and sometimes hungry because their family income had been spent in saloons and gambling houses.

Today, she felt, we do not often think of the heartbreak in these wild, booming towns. Although not always apparent, it was always there.

The San Juan communities were showing their age in the 1900s. After a mud-delayed, rough trip, George Backus arrived in Rico at 2:30 on a September 1909 morning. He finally located a rooming house, with the "worst bed I ever saw except one." Upon arising, he found Rico "a deserted, broken down mining camp." Another man explained its condition in these words, "too much hard work for low pay now"; the gilt had eroded from Rico. Buildings stood deserted, partly collapsed, broken windows blindly staring at a street on which few customers walked. Population figures coldly tell the story. From 1900-1910 Rico's population dropped by more than half, to 368, Sherman's from 176 to 18, Ironton's to 48, Red Mountain's to 26, and Lake City's slipped nearly half, to 405. These were the ones that still had residents; a score of once "promising" sites stood empty; no one hailed the traveler as he passed by.

Strater Hotel, Durango

Even with the specter of prohibition and the constant pressure of the WCTU hanging over it, the saloon flourished. Some of the "boys" in this Durango establishment look fairly rough. No women need enter here.

Western History Research Center, University of Wyoming

By 1900 many of the San Juan mining camps were showing their age. La Plata, in the canyon of the same name, had never really boomed and now was a decade past its best years. Log cabin architecture predominates.

A few managed to grow. Eureka climbed to 87, better than doubling its 1900 size. The earlier completion of the Silverton Northern Railroad to this point helped, as did a mining revival. In his last San Juan railroad building venture, Mears extended his Silverton Northern to Animas Forks in 1904 over his old road bed, then shot a short, unprofitable branch up Cunningham Gulch.[7] The steep grade to Animas Forks prevented much freight from being hauled up and thoroughly tested brakes coming down. Nor did the Gold Prince mine, on which he based most of his hopes for freight, prove up. Mears, the road and rail builder, had ended a generation of San Juan building; he was one of the few real old timers left.

For all the railroads, freight declined as mining retrenched. Mears even allowed the Red Mountain depot to be occupied rent free, if the person would "occasionally" clean around the building and give the still operating Silverton railroad a little space for baggage and other freight, "but this will put you to little or no inconvenience." Passenger service, generally tourists now, was the only bright spot in an otherwise gloomy picture.

From Creede to Ouray to Durango, tourism was picking up. Ouray boldly championed itself as "peerless Ouray, Most Picturesque Town in the World," not just in Colorado or the West. To fish, hunt, sightsee, take the "waters," experience the "wild frontier," even to ride the rails over some of Colorado's major engineering triumphs, the visitors came. Nicknamed the Rainbow Route, the trip from Silverton to Ouray was still extremely popular, more so on the Ouray than the Silverton end. Silverton, Mears commented, had a delightful summer climate and grand scenery, yet the locality remained unfamiliar to the tourist and offered limited accommodations. The Ouray to Ironton Park gap had never been mastered (people still dreamed of doing it); the railroads at either end stood at the mercy of the stages, bygone relics that lent charm to the trip. The railroad and stage companies split the ticket price, a constant source of annoyance and complaint. Petty squabbles also aggravated the feud, such as the stage's failure to deposit passengers at the depot. Caught between the Denver and Rio Grande and the stage, the Silverton Railroad fought to keep its fair share of the trade. The year 1901 seemed to have been a chaotic one for the Silverton: two stage lines were fighting over the right to carry passengers, the railroad itself was burdened with heavy expenses and light revenues, and the D&RG was stubbornly holding out for the Rainbow Route to be operated the way it preferred. Tourists finished the course, unaware of management's headaches in trying to insure that their trip was as pleasant as possible.[8]

The sightseers went home with postcards, snapshots (Kodak cameras were the rage), souvenir spoons, and other mementos. They also shipped 6½ tons of trout out of the Creede area by train in 1905. Their economic impact and importance grew; tourism emerged as a major industry, at least for certain communities.

The city fathers perceived the benefits and worked hard to stimulate tourism. Civic image was important morally, as discussed, and visually. Annoyed over a flood of posters, signs, and bills tacked up and plastered all over town, Telluride moved to stop the practice by making it unlawful without a license. Ouray's noxious dump was another source of irritation, a horrible stench "polluting and poisoning" the air, locals complained. In 1905 the council debated the issue for six months, to a draw, before cold weather moderated complaints and urgency. Cost considerations overrode those of health and appearance this time. They had more success in recommending that house numbers be affixed or replaced on homes. Lake City's marshal simply served notice on property owners who did not remove rubbish from their lots, to what results the minutes were silent.

The cost of governmental operations varied tremendously, indicating plainly the prosperity of the town. Ouray, in 1900-1901, appropriated $30,000, Silverton $48,000 for 1903-1904, and Telluride $28,000 for the year beginning April 1, 1902. Creede, a year later, budgeted only $5,800, Rico, in 1905, $4,800, and Lake City $5,400 in 1909. Of these six communities, Rico was struggling the most, evidence of the district's depression. Rico tried everything short of abolishing municipal government. Annual appropriations were pared, salaries

*What saloon did not have its painting?
This Telluride example is reportedly
one of the local madams, Audrey by
name. According to legend, she married the artist.*

sliced, twenty-year-old water bonds refunded, the office of marshal abolished (the town
clerk was authorized to act in his place), and services curtailed. To entice members for the
volunteer fire company, Rico provided free water to their homes. The "M&S Hose Co."
failed, nevertheless, and Rico was without fire protection until a new company was
organized.

More progressively, Ouray, Rico, and Creede passed ordinances granting telephone
companies the right to establish and maintain sytems. Silverton, Telluride, Lake City, Ophir,
and points in between also had phones, the Ophir exchange especially proud of its twenty-
four-hour service, a "distinct recognition of the importance of the place." The telephone
gained quick acceptance, as it had with the mining companies. So popular was it that the
Ouray Plaindealer warned patrons not to hog the line "talking nonsense."

Through it all municipal government kept functioning. It got no easier to keep city officials; salaries stayed low, complaints high, and jobs taxing. In 1900 Ouray had a miserable
time finding anyone who wanted to be dog catcher. It finally succeeded and paid him a flat
fee for every animal caught and impounded. (In Rico "Towser," "Bounce," "Trixie," "Bus-
ter," and "Headlight" continued to lope around town, delighting in their unlicensed and
uninhibited freedom.) The owner, when reclaiming the dog, paid for its room and board.

The dog catcher was not Ouray's most popular official. On the other hand, Telluride saluted a police magistrate who had served twenty-four years.[9] A new century brought new demands, but old problems continued to badger city fathers, including unpaid bills left behind as reminders of more prosperous times.

Those seemingly effulgent earlier years coaxed the first notable literary efforts from the San Juaners. The Ouray poet and ex-miner, Alfred King, wrote of them in this manner:

> A dream is the ghost of a fond delight
> An echo of former smiles and tears,
> Wafted to us on the wings of night
> From the silent bourne of the vanished years.

King, despite his blindness, published two collections of poetry, one containing his epic length "The Passing of the Storm," which recounted tales of miners' experiences. Mining engineer Frank Nason, former superintendent of the Silver Pick mine on Mt. Wilson, went home to Connecticut and published *The Blue Goose* and *To the End of the Trail*, novels with settings in the Telluride and San Miguel areas. Thomas Rickard published his reactions and impressions from a fall 1902 trip through the Telluride, Ouray, and Silverton districts, first serially in the *Engineering and Mining Journal*, then in book form, *Across the San Juan Mountains*. Of the three, only King stayed to write in the San Juans; the others simply used their experiences and the locales. Even so, it was significant that the San Juans and mining were considered subjects for these literary efforts.

The San Juaners themselves had more leisure time to read and contemplate those literary endeavors. Before moving away with his Camp Bird wealth, Thomas Walsh was determined to leave behind something of substance for his former neighbors. Perhaps emulating Andrew Carnegie, he provided the money to build the second floor of the Ouray City Hall, intended to be used as a public library. The Walsh Public Library contained 7,300 books by 1906 and patronage of fifty persons per day. Pay was not very high for the librarian and assistant; the city hall janitor received as much as both of them. Meanwhile, Silverton and Durango sought Carnegie funds for their libraries.[10]

As the first decade of the twentieth century drew to a close, the imprint of the nineteenth gradually faded. Old San Juaners might have yearned for what had passed; to the newcomer it was just King's dream. Upon the foundation of those early years, much was being built. The last lines of a verse of poetry appearing in the *Silverton Standard*, January 3, 1903, expressed it well:

> They builded better than they knew
> Those men now gone.

[1] *Lake City Times*, January 4, February 1, 1900, July 11 and 18, 1901. *San Miguel Examiner*, May 1, 1907, May 30 and August 8, 1908. Frank Hough Diary, December 31, 1900, Colorado Historical Society.

[2] Unless otherwise stated, the following composite was taken from: Hough Diaries, 1900-1901; Ernie Hoffman interviews, 1973; Carrie Craig Dyer Interview, *Pueblo Chieftain*, June 29, 1973; Norman Bawden Interview, *Silverton Standard and Miner*, September 22, 1972; and letters to the author from Martha Gibbs, Mathilde Moses, Margaret Miller, Mildred Ekman, Harriet Backus, Guy Emerson, and Alex Botkin; Gene Gressley (ed.), *Bostonians and Bullion* and Harriet Backus, *Tomboy Bride* (Boulder: Pruett Press, 1969). Madelin to Linquist, undated, copy in author's possession.

[3] *Thirteenth Census of the United States Abstract* (Washington: Government Printing Office, 1913), pages 596-611.

[4] Police Magistrate Docket, Rico. Police Magistrate Reports, Silverton. Lake City, Minutes of City Council Meetings, 1901. For politics see *Telluride Journal*, March-April 1903; *San Miguel Examiner*, April 1907; *Rico News-Sun*, April 1901; *Ouray Herald*, November 1906 and April 1907; *Creede Candle*, November 1908; *Silverton Standard*, April 1908.

[5] *San Miguel Examiner*, February 2 and June 8, 1907. *Ouray Plaindealer*, May 2, 1902. *Ouray Herald*, May 9 and 16, 1902. *Silverton Standard*, January 31, 1903. *Revised and Compiled Ordinances of the City of Telluride* (Telluride: Journal Publishing Company, 1909). Minutes of the City Councils of Creede, Telluride, and Lake City.

[6] James Batcheller to Harriet Backus, March 25, 1944, and c. May 1943. For the YMCA see *San Miguel Examiner*, November 1906-September 1909.

[7] Crum, *Three Little Lines*, pages 30-34. Silverton Northern Records, Mears papers. George to Harriet Backus, September 23, 1909. See footnote 3 for population.

[8] Silverton Railroad Records, Mears Papers. *Creede Candle*, 1906-1907. *Ouray Plaindealer*, December 30, 1904. *Sights, Places and Resorts in the Rockies* (Denver: Carson-Harper, 1901), pages 26-40.

[9] Lake City, Rico, Creede, Telluride, Silverton, and Ouray, Minutes of the Council Meetings, for various years between 1900-1909. *San Miguel Examiner*, February 16, 1907 and November 13, 1909. *Ouray Plaindealer*, January 20, 1905.

[10] Ouray, Minutes of the City Council, 1900, 1905, and 1906. Alfred King, *The Passing of the Storm* (New York: Fleming H. Revell Company, 1907), pages 128-129. *San Miguel Examiner*, May 30, 1908. T.A. Rickard, *Across the San Juan Mountains* (New York: Engineering and Mining Journal, 1903). Muriel S. Wolle's two delightful books, *Stampede to Timberline* and *Timberline Tailings*, have a wealth of San Juan stories.

B y the twentieth century's second decade the San Juans had filled out, like an adult ten years beyond youth. The freshness had faded, never to return. There would be no last hurrah, no last fling against relentlessly advancing time. Where once the miner stood master, now came the rancher and farmer, more mundane but more permanent. La Plata and Dolores counties were making the transition, not necessarily easily nor completely as yet. Mining had, for a long time, been secondary to farming in Rio Grande County; even this toehold virtually disap-

Yesterday's Shadow, Tomorrow's Unknown

peared. In Hinsdale County, mining collapsed; unfortunately, no other industry arose to fill the void. Only the core—San Miguel, San Juan, Ouray, and to a lesser degree Mineral— held out against the onrushing night of the mining era.

Population peaked and then melted away throughout the mining counties and communities. The tourist came to gawk at the relics of mining, wind battered and already being vandalized. Living in Creede in 1914, Annie Laurie Paddock, an elementary teacher, felt that the people "depended more on tourists," having "lost faith" that Creede's mining would come back. That loss of faith said so much about what had passed; it epitomized the difference between youth and age. Mining would go on, producing some solid, even spectacular, years, but something was lacking and never again would the whole San Juans experience the same intoxication.

In these last years before World War I ended a San Juan and an American era, the trends noted earlier, consolidation and corporate control, continued to dominate. The strikes and violence had taken their toll; union locals no longer openly competed for supremacy, although they were by no means dormant. The miner had become an industrial worker, his occupation stripped of whatever romanticism and individualism it had once possessed. Tramp miners came and went, while a core of faithful labored at each mine, some having worked for the company a decade or more. Job security outweighed other considerations. The West as a land of opportunity was as meaningless for these men as it was for the eastern factory worker. Words only, they offered nothing tangible for the eight-hour-a-day man, who was happy enough to have a pay check to take home. Many of his contemporaries were not so fortunate, as mining relentlessly receded, leaving fewer and fewer job opportunities.

Winter still took its toll and demanded caution, regardless of the century or the year. T.E. Frothingham made a December trip from the Tomboy mine over to Red Mountain in 1912. The slides were "splendid, if fearful," he wrote. While crossing a particularly dangerous slope, he heard a crack above his party. "We stood silent looking. Down it came. We could not evade anything anyway in the world. It started just ahead, the crack coming toward us. It was a wonderful wave came sweeping down." Finally with "diminishing seething," it stopped. Fortunately, the 600-foot-wide slide did not bury anyone, a trapped comrade was rescued, and they all continued cautiously on to the safety of the timber. "The word timber will always have a new meaning to me," confessed Frothingham. Reflecting on his adventure a few days later, he concluded that such things happen to "those men who dare these places the winter through." He closed with these bold words, "to me however it was a keen experience. I would not have missed it—as it came out—for a good deal."

In accounts of San Juan mining, certain words and phrases recur, indicating that Creede was not the only town losing faith. They are found in year-end summaries, mining journals, and even now and then in the local press, which struggled to promote a town's best image. "Operated in a small way," "nearly exhausted," "seen its best days," "comparatively inactive," "early big producer," "old mines coming into prominence" were a few of this genre. Even the last one, which might sound optimistic, merely described past glories and held out the hope that those days might return, if only . . . and that was a big *if*. Other signs of the same trend were mines that once employed large crews now being worked by ten to fifteen miners. The dumps of old mines, being worked or reworked to salvage low-grade ore thrown out earlier, gave insight into the actual conditions in a mining district. Perhaps this comment summarized conditions as well as they could be: "does not yield profit in proportion to difficulties encountered and risk." When a mine hit this point, it closed, and mines were closing throughout the San Juans. Now the local newspapers and residents lived on hopes: that capitalists and capital would ride to the rescue to underwrite exploration and development, that a new vein or ore pocket would be discovered, that the price of this or that metal might go up, thus encouraging low-grade ore production, or just that something encouraging might happen. While hope could maintain a district for a while, sooner or later the day of reckoning came.

Confounding the dismal trend were San Miguel's Smuggler-Union, Tomboy, and Liberty Bell. They continued to produce as they had previously, a record which the press took for granted and failed to cover as it once did (see table 1). With less flamboyance, Bulkeley Wells charted the Smuggler-Union's destiny, a property which he had long since returned to direct mining, rather than leasing. The Smuggler and Liberty Bell became embroiled in a nasty squabble in 1913 over trespass and ore removed from the former. The Liberty Bell lost the judgment and appeal and paid over $400,000 in damages. Fortunately, the mine was coming off two years of million-dollar-plus production, but the superintendent was so angry that for years he refused to speak to a friend who had served as a lawyer for the Smuggler. Errors in surveying and the question of where the vein apexed led to the trouble. The Tomboy, with much less fuss, maintained its profitable ways, purchasing the neighboring Montana claim in 1911 and thereby prolonging its life (the original Tomboy vein no longer was a major producer). The company pursued its modernization program, in 1914 adding a

Table 1.—San Juan Production, 1860-1914*

County	Gold	Silver	Lead	Zinc	Copper	Coal
San Miguel	$43,804,891	$21,579,554	$5,149,255	$738,558	$1,242,659	0
San Juan	20,438,444	17,208,265	10,489,022	861,967	5,872,759	0
Dolores	1,938,377	8,808,386	1,342,872	432,617	509,010	0
Ouray	32,167,763	26,051,774	6,052,912	30,650	2,897,804	0
La Plata	3,337,912	1,004,408	9,747	0	32,527	4,624,667 tons
Rio Grande	2,348,085	169,914	1,827	0	19,826	0
Hinsdale	1,422,615	4,411,112	3,884,250	55,823	384,097	0
Mineral	2,612,113	26,906,736	8,178,556	1,456,410	32,121	0

These production figures are the best estimates available. No actual total production figures will ever be known. Coal production is for the years 1892-1914. No other San Juan county listed coal mining for these years. Early reports for La Plata and other counties are sketchy. La Plata County averaged about 32,000 tons per year from 1885-1891.

* Sources: Henderson, *Mining in Colorado*, and Colorado Inspector of Coal Mines Annual Reports.

(above) Railroads never reached all the districts, so the ore wagons still rolled. This one is at Animas Forks.

(left) Uranium mining went through its first excitement just prior to World War I. The Primos Chemical Company was one of the major operators, showing that the industry still had a way to go before equaling the gold and silver mines. This mine was near Newmire.

cyanide mill with a 400-ton daily capacity. By 1913-1914, with San Miguel the third largest gold and silver producing county in Colorado, these three mines accounted for 90 percent of production.

The Camp Bird fared less well. Predictions of this mine's demise appeared as early as 1910, with forecasts that ore reserves were running out. Newspapers speculated about it, and a consulting engineer, J.E. Spurr, examined it and stated about development work, "it has opened up very little ore, and has in so far been somewhat disappointing." Production stayed high until the year ending June 30, 1912, after which it dropped by a million dollars to the $600,000 range. The Camp Bird was finished as a producer on the scale of the past decade. Spurr's estimate that continued exploration might result in new discoveries did not materialize. The low-grade ore reserves declined rapidly. Ouray had enough operating mines to remain a key San Juan district, although it dropped behind San Miguel.[1]

San Juan County's solid production allowed it to challenge Ouray for the number two position in the San Juans. Otto Mears, who, with partners, gained control of the Gold King mine at Gladstone and leased the Silverton, Gladstone, and Northerly Railroad running to it (purchasing it in 1915), finally brought this property and district into prominence. Just across the mountain, the Sunnyside mine near Eureka maintained steady production under the direction of the Terrys—father John H. and sons Joe and William. A fifty-niner who came to Silverton in the 1880s, Judge Terry, as he was known, had brought the Sunnyside through lean times to prosperity. The hard-working Terry deserves credit, along with Stoiber, for creating a method to concentrate low-grade ore, a process which gave new life and a brighter future to San Juan County. The Terrys acquired more than their share of stories and legends, and developed their mine and mill into a showcase of the district.

The Silver Lake mine slumped in the meantime and was turned over to lessees, who maintained operations at levels reduced from the days of Stoiber. The Iowa-Tiger, an extension of the Silver Lake vein, picked up the slack from its more illustrious neighbor for a couple of years, keeping activity at a high pace in the Arrastra Basin region, where the Little Giant had incited mining fever so many years before.

Zinc, maligned and detested by owners because of the penalty assessed against it by smelters, acquired new luster during these years, when new methods of treatment made it profitable to smelt and thus mine. Although the great boom hit Lake County, the San Juans produced their share, primarily in San Miguel, San Juan, and Dolores counties. Rico revived for one short year in 1913, thanks to zinc, lead, and copper. With plenty of buildings standing vacant and the mines already open, there existed no need for an old time rush to rejuvenate the town. Tungsten briefly excited Silverton. The growing demand for steel, which tungsten was used to harden, opened a market. The whole state was caught up in the excitement, which centered on Boulder County. Silverton's deposits proved too small to be of much significance.

Telluride found itself on the edge of uranium country; Placerville and Newmire, where mill and mines were located, sat just down the river. Many claims were filed there and farther west, and small-time speculation fanned interest. As in any other new district, miners found poor roads, long distances for transport, and reduction problems. Unlike the market for gold and silver, this one was limited, further handicapping the industry; as early as 1912, an oversupply curtailed mining. Burros and wagons hauled freight out to the mines and brought back the ore; the Paradox Valley continued to be the center of activity. The Primos Chemical Company, which operated a concentration plant at Newmire, tried using "auto" trucks to transport its ore; "bad roads" foiled the experiment and compelled the company to resort to wagons. The search for uranium spread to other western slope areas, and the spring of 1914 found many prospectors in the field, staking claims on likely looking rocks, many not exactly sure what they were seeking.

The coming of World War I dashed their expectations. Europe had been the chief

Nature had to be respected, no matter what year. This Telluride flood in August 1914 did a great deal of damage. Mary Waggoner witnessed it and wrote a friend, ". . . such desolation you cannot imagine."

Few skiers today would appear in such outfits as these adventurous folks outside of Animas Forks. Yet it cannot be said that the crude equipment in any way lessened their fun.

market, and it now had other needs to fill. Uranium purchases dropped to nearly nothing. To make matters worse, recent production far outstripped demand and left a glutted market. Mines closed, miners were laid off, prospecting ceased, and a peacefulness returned to the plateaus, scarred now by man's passage. The first uranium boom had run its course.[2]

World War I did more than collapse the uranium market; it adversely affected the entire San Juan region. The outbreak of war immediately checked all British-backed expansion programs and stopped the flow of money from England. Many British companies inaugurated a policy of selling their American investments and bringing home the funds to be used there in the war effort. Emigrant miners, especially those from eastern Europe, confronted divided loyalties once the war began in earnest. Nor did the San Juan mining communities escape the superpatriotism and bigotry that characterized other parts of the United States. War pressures opened new markets with high prices and partly closed others, leaving miners, for the first time, at the mercy of a world-at-war market. Shortages came eventually, along with meatless and wheatless days. Mines near exhaustion or in low-grade ore closed, as wartime demands left them adrift, or the postwar depression dashed their final hopes. The boys marched off to war in 1917 and 1918, draining some of the surplus miners from the district. A few went "over there" and when they got home, the San Juan world they had known in 1914 had vanished. Those vanished yesterdays soon seemed sunnier, more secure, and more exciting than the present, as nostalgia took hold. Only Telluride would keep the prewar pace into the 1920s.

Annie Laurie Paddock, who remembered Creede as a town that depended as much, if not more, on tourists than mining, did not find it an exciting place for a young, unmarried woman. Amusements were few, as were young men. Given a choice between living in a home with electric lights or a bathtub, she resolutely chose the latter, only to discover that the water turned cold by the time it was pumped into the tub. Shopping was easiest in the pages of Sears's and Ward's catalogues, local merchants having little to offer. Specialized businesses had left, a sure sign of a camp's decline. What a difference twenty years had made in Creede!

Women teachers were expected to adhere to certain standards, no change from earlier times; Annie Paddock taught Sunday school and became a member of the local literary society, which studied history in 1914. One vividly remembered session included the statement that there would "never be any more wars." Bridge was popular (whist captured adherents in other camps), and a night-time stroll to a party along Creede's unlighted streets was an adventure in itself. Concerned as a teacher, she found the city government scarcely interested in education and the school without many books. To teach the would-be scholars, who did not seem to have changed their mischievous ways over the years, Annie Paddock was paid $75 and boarded on her own.

Creede's condition reflected the plight of other San Juan mining towns and camps. Obviously, some were in worse shape than this. Their newspapers and businesses were gone or going, population had declined, and their mines had quieted to a faint whisper of their previous activity. The remaining newspapers resembled their counterparts in other towns of similar size, having lost much of the individualism and zest that gave their predecessors a frontier-mining flavor. Mining columns and local news still appeared, but routine developments were now spiced with only an occasional flash of some small sensation. Editors never gave up hope, however, and continued to forecast better times, or, as in the case of Silverton and Telluride, a lengthy continuation of present conditions. National and international news received fairly thorough coverage, probably because a scarcity of "real" local news provided extra column space. Annie Paddock and her friends read *Creede Candle* editorials that commented on everything from Creede to national issues, with even an occasional blast at something of an international nature. Grain and stock market quotations and American and National League baseball scores occupied space once reserved for mining news. The quality of writing stayed surprisingly high, considering the conditions of the communities.

Photographs and "modern" advertising methods gave the papers a much slicker appearance, and reliance on outside news sources made them all seem more alike than in Dave Day's era. Editors assumed that readers wanted stories of the unusual, sensational, and maudlin, because many of those kinds were featured. As clearly as any source, the newspaper mirrored the changed San Juan situation, portraying the ties to the past and charting the paths to the future.

Election time always heated tempers and emphasized partisanship. In 1912 the Democrats held sway, Woodrow Wilson capturing the San Juans and Colorado. Of the state legislative delegation that year, San Juan Republicans managed to win only one seat. Telluride continued its socialistic bent, Eugene Debs garnering just over 100 votes, compared to the winning Democrat's 900-plus; the Socialist vote percentage declined significantly from earlier elections. A small core of loyalists had obviously remained faithful since the days of the strike. In 1914 the San Juans voted wet, Colorado dry. The *San Juan Examiner* put it this way, "the prohibition wave has struck us, and incidentally engulfed us." The editor hoped the law would soon become obnoxious, then be repealed or modified. He who lived on such hope soon found his favorite saloon only a memory.

The best way to advertise the San Juans continued to be perplexing. Some individuals fretted that the region was neglected because of its long distance from "Denver and other central points." Silverton's Commercial Club organized a special publicity committee, which sent out articles to leading newspapers, "so written as to advertise the resources of the county." It also toyed with the idea of creating a lantern slide presentation and inviting newspapermen to come to Silverton to "show them a time." In 1913 the club published pamphlets, supported good roads, tried to get a railroad into the San Juans from the south (to break the D&RG monopoly), and conducted a letter-writing campaign in praise of San Juan County. By then the club had rooms where "social features" proved a great success, affording businessmen "much needed recreation."[3]

The *Silverton Weekly Miner*, October 25, 1912, sounded a familiar warning when it pointed out that no town could "grow whose people are untrue to it." "Support the town and its merchants" was the paper's credo. Anyone who destroyed confidence in his town was "an enemy to progress and a traitor." A year before, the *Creede Candle* put it this way, "for God's sake quit chewing the rag [consigning Creede to the boneyard] around the streets like an old fish-wife." The old attitudes, the old worries refused to die.

The Silverton Commercial Club was not alone in advocating better roads and another railroad outlet. Concern was voiced over the high freight rates for ore, rates established when the mines produced high grade. Lower rates, proponents reasoned, meant more mines could operate and more freight be made available to the railroad company. Complaints over high rates beyond Durango drew a Denver and Rio Grande response that the railroad was protecting the American Smelting and Refining Company's Durango plant. Appalled, the Commercial Club wrote back, "in discriminating in favor of this company, you are holding down the production of this camp and cheating yourselves out of big tonnage." The Club could not tolerate that kind of logic and finally exploded, "We do not think that the railroads have the right to protect one industry to the detriment of a whole community." A long line of American reformers fighting to curb the railroads' power understood such feelings.

Better roads were one way to curtail the power of the D&RG; already automobiles were making an appearance, and trucks, as mentioned, were being tested as ore haulers. As far back as one can go, San Juaners had complained about their roads. In the 1890s bicyclers had organized and fought for good roads; now the cry became more general. Improved roads facilitated transportation and encouraged tourists. The novelty of the automobile attracted press attention upon its arrival. In 1910 the first car over Stoney Pass (in only slightly better shape than it had been when cursed in the 1870s) took five days to accomplish the feat and had to be helped along by horses. Silverton turned out to welcome this symbol of progress, which toured on to Ouray, again with horses' aid. Within a year a local resident was driving his own car down Greene Street. By 1914 the Creede paper reported tourists driving all the way from Kansas City to relax amid the beauties of Mineral County. Local roads, the paper proudly announced, had merited praise from these adventurous travelers, who encountered muddy roads elsewhere. Other camps hurried to put their roads in equally good condition to attract their own tourists and grumbled when the county seemed slow to respond. Ouray sent delegates to a "good roads" convention in Pueblo in January 1912, who accomplished little for the region they represented. Eastern slope interests proved too strong; once again their weakening economic and political position in comparison to the eastern slope was brought home to San Juaners.

One can imagine the horses' reaction when they first encountered this noisy new beast on the road. Numerous were the rearings and runaways when horse met car, with attendant accidents and imminent danger to drivers, riders, and bystanders. The problem seemed to be one for the city councils to resolve. Some (Ouray, for instance) tried imposing speed limits.

The councils now confronted some other unusual issues. An audit uncovered the startling fact that Telluride's treasury was short $36,000, and the treasurer precipitately departed for safer climes. A four-month hassle with the man's bonding company finally produced a settlement of $33,000, while the public followed the tracking of the fugitive via the press. Lake City suffered through a measles epidemic in 1911; the one doctor who would accept the position of city health officer did so only after the council agreed to support him and pay his fees. His best efforts did not prevent the epidemic from running its course, amid complaints and countercomplaints over his demands and lack of municipal backing. After an outbreak of vandalism, Lake City offered a $10 reward for information leading to the arrest and conviction of anyone interfering with electric lights, wires, or globes. Arrested suspects, if any, were placed in the brand new steel jail cells ordered from St. Louis. With high hopes of a profitable new revenue source, Rico placed a license fee on "moving picture shows," which caused much grumbling and led to a reduction of the fee to $2.50 per month.

Ouray kept after its rubbish problems, as thoughtless residents continued to dump trash in any available spot. A new problem arose over gasoline pumps and fuel storage, resulting in an ordinance prescribing acceptable locations and methods. The marshal encountered a minor problem with ice peddlers, who dumped sawdust from the ice on the sidewalk, which eventually filled the gutters. The council, through the marshal, advised peddlers to clean their ice before unloading. The Ouray city fathers investigated the possibility of purchasing a police call system, but on "account of the financial condition" of the town, laid the matter on the table. Those were the problems bedeviling these towns in the century's second decade.[4] Except for Telluride all were in a financial crunch, forcing them to reduce services and eliminate municipal positions.

Other unique or unusual phenomena enlivened the San Juaners' generally routine life. Halley's comet aroused considerable interest in 1910 when it illuminated the night sky. Movies, of course, were an attention-getter. For only 10¢, the Creede motion picture theatre promised three reels at every show. And the "pathetic and thrilling" "When the Earth Trembled" enthralled Silvertonians in August 1914. Basketball caught the public fancy as well; it could be played indoors and required fewer players, making it ideally suited for the local school and town teams. The largest schools fielded both boys' and girls' teams. Silverton, Ouray, and Telluride had probably the best school systems and were justifiably proud of them. Telluride, for example, employed fourteen elementary and high school teachers in

As the new century entered its second decade, the red lights started to go out. This is the inside of a crib, or one-woman establishment, at Telluride. Few pictures like this are known to exist.

The era was now over. As David Lavender wrote years later, "something of the directness, the simplicity has disappeared. You can't mine paper; a pay check isn't the same as seeing your own gold in your own hand. Even so the pride persists. The miner goes to his wet, lonesome sunless trade with his head up; . . ."

1910, and the *Journal* devoted almost an entire issue to the schools and the high school graduating class that June.

The local newspaper occasionally took upon itself the task of informing readers on matters of etiquette. The *Creede Candle*, July 29, 1911, recommended that the telephone operator be treated in a kindly manner. Be as courteous to central, the editor urged, as you would to any lady with whom you were speaking. "Central is very human and a courteous word is not wasted upon her."

Basketball, automobiles, movies—the San Juans were changing, and so was mining. Mining's diminishing role in Colorado was brought home to the San Juans when Bulkeley Wells, president of the Colorado Metal Mining Association in 1914, gave a speech in Silverton. His main complaint centered on the alleged discrimination practiced by the legislature against mining properties by virtue of a recently passed tax law. His special ire focused on the legislative members from the farming districts, whose sympathy did not lie with mining. The agricultural and urban districts had been steadily gaining power at the expense of the mining districts, a fact now becoming abundantly clear on several fronts. Until the past decade, mining had had pretty much its own way; now that dominance was eroding and the adjustment to new conditions proved difficult.

In August 1914, news of the outbreak of war reached the San Juans. It could not have come as a complete surprise, since the press had adequately reported the heightening tensions. For those San Juaners with family and emotional ties to the old country, the news was upsetting; for many eastern Europeans and Germans, it was heartbreaking, as they watched America align itself against their motherlands. These most recent arrivals were well down the path that immigrants before them had traveled: assimilation. More and more eastern European names appeared as businessmen and leasers of mines; the doors of the American dream were opening for them and even more so for their children.

Early press coverage was straightforward reporting of events, the recent Telluride flood with its loss of life and property competing with the war news for space. By the middle of August the editors started analyzing. The *Creede Candle* innocently wrote, in reply to Germans who considered its press reporting unfair, "We have the notion that the horrors of war lie not so much in the spreading of inauthentic reports as in deadly bullets." The situation, the papers agreed, was serious, and thousands would be killed unless people and nations came to their senses. By the end of the month, photographs from the front, battle stories, and war news columns had become regular features; the San Juaner knew what was happening nearly as well as any other American.[5] Though the aforementioned immigrants and some other San Juaners anguished over the international scene, many probably reflected the attitude that Annie Paddock reported in Creede. There was little concern in the town at the war's inception—there were not enough people left. A few boys would volunteer, but the people were mostly of American background and did not worry about Europe's problems. No doubt they were happy to be in the San Juans, far from the battle.

As if to recapture lost youth and hold at bay the rush of time, two short mining excitements provided diversion in 1913-1914. Ironically, Summitville made news again, just as it had at the start in the 1870s. To the south of Summitville, at Platoro, rich surface gold ore was found, the "new Cripple Creek." The area was neither rich nor new—prospectors had been there in the 1880s and 1890s, cursing the refractory ore. Nor did ore averaging $10 to the ton long remind them of Cripple Creek. Cars and trucks hauled expectant rushers in and disappointed ones out; by mid-1914 the "boom was going the other way," as they say.

The year 1914 turned the attention of those who were intoxicated by new mining discoveries to Cave Basin, located on the La Plata and Hinsdale county lines above present Vallecito reservoir. When word leaked out that spring of rich float and high-grade ore, men traveled on snowshoes to reach the spot. Later an "automobile stage" eased the way. The *Durango Semi-Weekly Herald*, March 26, predicted that the district would be "one of the best in the state." Forecasts that 1,000 men would crowd into the basin, followed by investors, missed the mark considerably. Cave Basin had many claims, but small production, and soon only the sound of annual development work disturbed nature's serenity.[6]

As last flings, these proved pathetic reminders of a lost youth. In them, though, the whole saga of the San Juans passed once more, mingling the old and new, the excitement, the disappointment, the dreams, and the reality. A new generation could experience, if ever so diluted, the thrill of a mining rush. The opportunity would not be theirs again. Alfred King, musing on this years before, eventually wrote:

> I live, I move, I know not how, nor why,
>   Float as a transient bubble on the air,
> As fades the eventide I too, must die;
>   I came, I know not whence: I journey, where?

[1] J.E.Spurr, Report on the Camp Bird, 1910, J.E.Spurr Collection, University of Wyoming. *Engineering and Mining Journal*, 1910-1914. Annie L. Paddock Interview, July 18, 1974.

[2] *San Miguel Examiner*, 1912. *Engineering and Mining Journal*, 1911-January 1915. *Mineral Resources of the United States for 1912* (Washington: Government Printing Office, 1913), volume I, pages 1004-1007. Bulkeley Wells later was involved in uranium claims.

[3] Silverton Commerical Club papers, Searcy papers, Western Historical Collections. Paddock Interview. For the newspapers, see the *Creede Candle*, *Silverton Weekly Miner*, *Silverton Standard*, *Ouray Herald*, *Lake City Times*, and *Telluride Journal*.

[4] Creede, Minutes of the City Council, 1910-1914. Telluride, Minutes of the City Council, 1910-1913. Ouray, Minutes, 1911-1913. Rico, Minute Book, 1913-14. Lake City, Minutes of the City Council Meetings, 1910-1914.

[5] *Silverton Weekly Miner*, August 7-21, 1914. *Creede Candle*, August 1-22, 1914. *Lake City Times*, August 13-September 10, 1914.

[6] *Engineering and Mining Journal*, 1913-1914. *Mineral Resources of the United States*, 1912, 1913, 1914, and 1917. *Durango Semi-Weekly Herald*, 1914, had very little on Cave Basin, a commentary on the changing situation.

Index